Birkhäuser

Tessa Morrison

Isaac Newton's Temple of Solomon and his Reconstruction of Sacred Architecture

Birkhäuser

Tessa Morrison
The University of Newcastle Callaghan
The School of Architecture and Built Environment
Newcastle, Callaghan
Australia
Tessa.Morrison@newcastle.edu.au

ISBN 978-3-0348-0045-7 e-ISBN 978-3-0348-0046-4
DOI 10.1007/978-3-0348-0046-4

Cover design: deblik, Berlin

Printed on acid-free paper

Springer Basel AG is part of Springer Science+Business Media

www.birkhauser-science.com

Foreword

Architecture is certainly not a field that comes to mind when we think of Isaac Newton. This is precisely why this present volume is so important. It not only shows us a little known side of one of the greatest minds in human history, but also helps us understand entire areas of inquiry that have fallen into oblivion, those of "chronology" and "ancient wisdom".

Because we are all intelligent beings, intelligence itself is of interest to us. Genius, which is superlative intelligence, is greatly interesting to us. But because genius is not always accessible, myth takes over. This is as true for historical figures like Newton and Leonardo da Vinci, as it is for more recent geniuses, like Einstein: the scientist is obscured by the myth. In the case of Leonardo, the myth grew to such proportions that the tendency is to credit Leonardo with the invention or prototype for almost everything, in spite of the fact that in many cases careful study of his notebooks demonstrates that, even when he correctly observed the phenomena he was studying, he drew the wrong conclusions, or when, in the attempt to build his "inventions", they are discovered to have irremediable flaws. The case of Newton is just the opposite. Newton's myth was forged by neglecting or denying the activities that actually occupied the largest part of his intellectual life. Recent interventions have aimed at restoring to Newton the parts of this work that earlier periods had deemed unimportant or even scandalous. This present work falls in that category.

Uncovering and making accessible the work of scientists and scholars of the past is much more difficult than it might appear to be. Newton himself knew that: had these challenges been easy, he needn't have spent a lifetime working on ideas surrounding chronology, ancient wisdom or Solomon's Temple.

One of the obstacles is language. In order to interpret the biblical passages he was interested in, Newton had to grapple with the Hebrew. But for many of today's scholars, a Latin text presents equal difficulties. Latin was, of course, the language of scholars. Not having had an education of the kind reserved to young noblemen, Leonardo da Vinci found that many mathematical and scientific treatises remained closed to him because of the Latin, but they are equally closed to today's scholars. In our present age, knowledge of Latin is limited almost exclusively to specialists. Even Newton recognised the limits of Latin. He himself was interested in the creation of an international language that could more accurately interpret the prophets that concerned him.

However, knowledge of the language is only one of the requisites for accurate translation; the other is knowledge of the subject. All translations are essentially interpretations. If the translator is not knowledgeable in the field, we are likely to find every genre of mistake.

Another obstacle to retracing lost ideas is related to the cultural changes that separate one epoch from another. According to one theory, we can never completely understand an earlier age, because we must inevitably filter what we read about that age through our own cultural alignment, that is, it is impossible for us to "unlearn" what we know so that we can approach the ideas that came before us, and therefore all of what we think about earlier ages is contaminated by our own. If we carry that to its logical extreme, of course, it makes no sense to study history because we can't know what ideas meant in an earlier age. This would doom us to a perpetual present, flattening out our experience. But we can't go to the other extreme either, taking it for granted that there are no cultural gaps. In the case of the seventeenth century, for instance, it is almost impossible for us to grasp how thoroughly Christianity permeated all aspects of culture and society. But in order to understand Newton, we must try to understand what that must be like, even if we are all the more surprised that religion would have such an overwhelming influence on the very mind we hold to be the paradigm of scientific thinking.

Culture also determines what areas of inquiry are deemed to be of the greatest importance. Our scientific age nods in agreement with Newton's contemplation of mathematics and physics, but finds it harder to countenance inordinate amounts of time and effort given to alchemy and interpretation of the prophets. We apparently agree with Newton that history is important, but we may disagree with just what about it is important. Newton concerned himself with "chronology", that is, the science of computing time or periods of time (not to be confused with measuring time), an area of inquiry that began to fade around the turn of the eighteenth century. In an age that was famous for its scientific disputes, Newton was perhaps the most polemic figure of all. His dispute with Leibniz over the discovery of calculus is legendary. His efforts to take revenge on Robert Hooke and consign him to oblivion are also well known. So it comes as no surprise to read that his results in "chronology" were hotly contested as well.

By now, however, you can see how many layers have to be gone through to reach some kind of understanding of Newton and his more esoteric interests. Tessa Morrison may well be the only scholar knowledgeable enough in both the Latin and this particular material to make her way through it. In having made the effort, she has given us back a part of Newton that we were seriously in danger of losing altogether.

If this book were merely a translation, however, it would be rather a dry academic exercise. What makes it especially significant is the commentary, which inserts Newton's manuscript into its proper context within the ongoing discussion about relationships between architecture and mathematics. While the Bible does contain some information about measurements, it is not one of the most precise documents in this respect. It is fascinating to learn that Newton was at pains to determine the exact length of the cubit. Even more fascinating is that Newton studied Vitruvius and derived his own Vitruvian man. Thus Newton is given his rightful place in the tradition of architectural thinkers such as Alberti and Leonardo. Further, his criticism and revision of Villalpanda's reconstruction of Solomon's temple shows him to be an acute architectural analyst.

As a translator and commentator, Dr. Morrison does a remarkable job of keeping herself off the page. But Newton scholarship, as well as that on architecture and mathematics, is greatly enriched by efforts such as these, and we can be grateful that she has dedicated her time and attention to bring this present work to our notice.

Editor in Chief, *Nexus Network Journal* Kim Williams
Turin, Italy

Acknowledgements

I wish to express my gratitude to my many friends that have offered support in the writing of this book, in particular Andrew Stephenson and Julian Poulter for their suggestions and to Ning Gu and Josephine Vaughan for their very useful advice on the 3D modelling program ArchiCAD. In this project, I have used a wide range of Newton's unpublished manuscripts which have been generously made available by many libraries. I would particularly like to thank the Babson College Library for making public an excellent reproduction of Babson MS 0434 on their website. To view the reproduction of the manuscript see http://www3.babson.edu/Archives/newton_collection/MS434C.pdf. Also thanks to The Newton Project at the University of Sussex for making many manuscripts public that would have been formally unobtainable. Furthermore, I wish to thank Kim Williams who provided a personal preface for this work; also to my husband John for his continuous, invaluable support and assistance which I could not do without.

Callaghan, NSW, Australia Tessa Morrison

Contents

1 **Introduction** 1
The Unpublished Manuscripts of Isaac Newton 3
Argument from Design 5
Structure of This Book 10
Conclusion 11

2 **Chronology, *Prisca Sapientia* and the Temple** 13
The Short Chronicle 14
The Chronology of Ancient Kingdoms Amended 17
Principia and *Prisca Sapientia* 22
Conclusion 27

3 **Prophesy and the Temple** 29
Influences 29
The Language of the Prophets 32
The Practices of the Prytanæa and the Temple 35
Is Ezekiel's Vision of the Temple the Same as Solomon's Temple? 39
Conclusion 41

4 **Prytanæum and the Floor Plan of the Temple of Solomon** 43
The Prytanæum as the Frame of the Universe 43
The Prytanæum or Temple as Microcosm of the Macrocosm 46
Stonehenge as Prytanæum 56
Maimonides' Floor Plan of the Temple 59
Conclusion 61

5 **The Temple Measurements and the Sacred Cubit** 63
Overview of Ancient Measurements 63
John Greaves and Egyptian Measurements 65
Newton's Interpretation of the Hebrew Cubit 65
Conclusion 71

6 An Overview of the Contents and the Source of Babson MS 0434 73
Overview of Babson 0434 73
The Sources of Babson 0434 75
Conclusion 77

7 Reconstruction 79
The Differences in the Images of the Temple in the *Chronology* and Babson MS 0434 79
The Differences Between the Description of the Temple in the *Chronology* and in Babson MS 0434 82
The Description(s) of the Temple in Babson MS 0434 84
The Reconstruction 89
Conclusion 93
Transcription of the English Annotation in Fig. 7.4 93

8 Conclusion 101

Translation of Babson MS 0434 105

Notes 157

Bibliography 175

Index 183

List of Figures

Fig. 3.1 Mede's floor plan or "plot" of the temple (Drawn by the author from Mede, 1643, part 2, p. 6) 30
Fig. 4.1 Villalpando's plan for the layout of the tribes of Israel's camp around the Tabernacle (Drawn by the author from Villalpando and Prado, 1604, vol. 2, p. 467.) 47
Fig. 4.2 Villalpando's astrological arrangement for the plan of Solomon's temple (Drawn by the author from Villalpando and Prado, 1604, vol. 2, p. 470.) 49
Fig. 4.3 Villalpando: A single colonnade and the resemblances to the division of the human stature (Drawn by the author from Villalpando and Prado, 1604, vol. 2, p. 472.) 50
Fig. 4.4 Montano's reconstruction of the Temple of Solomon, from his "Exemplar" in volume eight of the Antwerp Polyglot (Drawn by the author from Zur Shalev (2003), p. 64, with kind permission.) . . 52
Fig. 4.5 Claude Perrault's floor plan of the Temple from Louis Compiègne de Veil's translation, from Hebrew into Latin, of *The Code of Maimonides, the Mishneh Torah* originally published in 1678 (Drawn by the author from Louis Compiegne de Veil, 1683, unpaginated.) 53
Fig. 4.6 Constantin L'Empereur's floor plan of the Temple from Guglielmus Surenbusius; Mishnah sive Legum Mischnicarum liber qui inscribitur Ordo Sacrorum..., 1702 (Drawn by the author from James Stevens Curl, 1991, p. 89, with kind permission of Professor James Stevens Curl.) 54
Fig. 4.7 Villalpando's reconstruction of the temple (Drawn by the author from Villalpando and Prado, 1604, vol. 2, unpaginated.) 55
Fig. 4.8 Inigo Jones' architectonical scheme for Stonehenge (Drawn by the author from Inigo Jones, 1655, pp. 58 and 59.) 57
Fig. 4.9 Vitruvius' plan of the theatre (Drawn by the author from Vitruvius, 1960, p. 147 with kind permission.). 58
Fig. 5.1 Greek fourth century BC stone engraved with standard measurements (Drawn by the author at Piraeus Archaeological Museum in Greece) 64

Fig. 5.2 The Newtonian man (Drawn by author from Newton's description in Isaac Newton, 1737) 69
Fig. 7.1 The floor plan of the Temple precinct published in the *Chronology* in 1728 (Drawn by the author from Isaac Newton, *The Chronology of Ancient Kingdoms Amended* (Dublin, 1782), p. 346.) 80
Fig. 7.2 The floor plan of the Temple published in the *Chronology* in 1728 (Drawn by the author from Isaac Newton, *The Chronology of Ancient Kingdoms Amended* (Dublin, 1782) p. 346.) 81
Fig. 7.3 Floor plan of the cloister under the chambers published in the *Chronology* in 1728 (Drawn by the author from Isaac Newton, *The Chronology of Ancient Kingdoms Amended* (Dublin, 1782) p. 346.) 82
Fig. 7.4 The Temple Precinct as drawn by Newton in Babson MS 0434 (From Isaac Newton, *Introduction to the Lexicon of the Prophets, Part Two: About the Appearance of the Jewish Temple* (Babson Ms 0434) (unpublished manuscript, Babson College) fol. 9r, with kind permission of The Huntington Library.) 85
Fig. 7.5 The final floor plan of Babson MS 0434 (Drawn by the author from Isaac Newton's description in Babson Ms 0434.)ABCD: The Separate Place that contains the Temple
DCEF: The Court of the Priests or the Inner Court
G: The Altar; H: The kitchen of the priest and where the sacrifices were prepared; I: The Temple; KADJ and BLMC: The chambers of the higher ranking priests; JDFECMNO: The chambers of the priests; P: The gates; Q: The kitchens of the people and the stairs to the upper chambers; RSTUVWXY: The chambers of the people of the Outer court 88
Fig. 7.6 Elevations of the Outer Eastern Court Entrance (Drawn by the author from Isaac Newton's description in Babson Ms 0434.) 89
Fig. 7.7 Elevations of the Outer Northern Court Entrance (Drawn by the author from Isaac Newton's description in Babson Ms 0434.) 89
Fig. 7.8 Elevations of the Inner Eastern Court Entrance (Drawn by the author from Isaac Newton's description in Babson Ms 0434.) 90
Fig. 7.9 Elevations of the Inner Northern Court Entrance (Drawn by the author from Isaac Newton's description in Babson Ms 0434.) 91
Fig. 7.10 Elevations of the Temple (Drawn by the author from Isaac Newton's description in Babson Ms 0434.) 92
Fig. 7.11 Sections of the buildings of the Priests (Drawn by the author from Isaac Newton's description in Babson Ms 0434.). 93
Fig. 7.12 The following colour plates represent a walk through the Temple precinct (Drawn by the author to describe the path of the coloured images) 94

Fig. 8.1 The columns represent the number of words written by Newton from the 1660s when he began his studies at the University of Cambridge. Not represented in this table are Newton's Mint writings; there are 916 surviving Mint documents, which date from the mid-1690s to the last years of his life. These are receipts, instructions, lists of obligations and general correspondence and as such do not constitute a body of research therefore they have been excluded 102

List of Tables

Table 2.1 The contents of the drafted scholium 24
Table 5.1 Measurement of the Great Pyramid as given by Greaves and the ancient source 67

List of Plates

Plate 1	Birds-eye perspective of the Temple Mount	95
Plate 2	Birds-eye view of the Temple and its surrounding courts	95
Plate 3	The beginning of the walk through the Temple precinct and it begins in front of the Altar and the Temple	96
Plate 4	The first chamber of the Temple, decorated with cherubim and palm trees and the back of this chamber is the stairs to the Holy of Holies	96
Plate 5	The most sacred part of the Temple, the Holy of Holies, which contains the Ark of the Covenant. This room is only entered once a year by the high priest.	97
Plate 6	A section of the two chambers of the Temple	97
Plate 7	The Temple looking at the inner court.	98
Plate 8	The outer court turning right	98
Plate 9	The corner in the outer court	99
Plate 10	Turn back to the eastern gate.	99
Plate 11	A section of the eastern gate. A high priest is entering at the point at which Ezekiel first entered the Temple precinct.	100

Chapter 1
Introduction

In Michael White's biography of Isaac Newton he states, "According to a list of the most influential people in history, *The 100*, Isaac Newton ranks number 2 – after Muhammad and ahead of Jesus Christ".[1] An extraordinary statement written at the close of the twentieth century; a century of turmoil and rapid technological change, a century in which scientific changes and development have brought about the contrast of extreme misery and poverty, with massive prosperity and wealth. Moreover, in 1999, *The Sunday Times* named Newton its "Man of the Millennium".[2] For a scientist who died in the early eighteenth century to be considered so influential despite the duration of time and change, not only in science but also in attitude and thought, is unique. Of Newton and his time, Albert Einstein claimed that

> Newton's age has long since passed through the sieve of oblivion, the doubtful striving and suffering of his generation has vanished from our ken; the works of some few great thinkers and artists have remained, to delight and ennoble us and those who come after us. Newton's discoveries have passed into the stock of accepted knowledge (Einstein, 1952).[3] It was, however, Albert Einstein who demonstrated that the science of Newton was insufficient to describe the quantum world of sub-atomic particles. There is no doubt that Newton did play a most significant role in the history of science and mathematics. His achievements alone define him as one of the greatest scientific geniuses in history, but the myths that surround his memory redefine him not as a great mathematician or physicist but as an inspired dreamer.

In the collected memory of the public, Einstein is associated with the mathematical equation $E = mc^2$. Although most people would not understand the full implication of this equation, at least Einstein, a scientist, is associated with an important contribution to science. Newton could equally be associated with the mathematical expression $1/r^2$ – the relationship of the inverse-square law – in which Newton suggested that the force of gravity that acts between any two objects is inversely proportional to the square of the distance between the centres of these objects. Yet Newton is more likely to be associated with an apple tree, the (mythical) inspiration of his genius, rather than the ingenious work that resulted from that inspiration.

The image of the inspired genius, devoutly religious and working in isolation for the betterment of mankind, has been propagated throughout the centuries.[4] His early biographers such as William Stukeley[5] in the eighteenth century and David Brewster[6] from the nineteenth century, and many others, often glossed over manuscripts, letters and events that did not fit into this image of the pristine inspired dreamer.

T. Morrison, *Isaac Newton's Temple of Solomon and his Reconstruction of Sacred Architecture*, DOI 10.1007/978-3-0348-0046-4_1, © Springer Basel AG 2011

This may have been an image that the ageing Newton wanted to project, for the apple story originated from Newton himself, late in his lifetime. In William Stukeley's *Memoirs of Sir Isaac Newton's Life,* he related how, in 1726, Newton told him that one warm day sitting under an apple tree at his home Woolsthorpe in the summer of 1665

> The notion of gravitation came into his mind. It was occasioned by the fall of an apple, as he sat in a contemplative mood. Why should that apple always descend perpendicularly to the ground, thought he to himself. Why should it not go sideways or upwards, but constantly to the earths centre? Assuredly, the reason is, that the earth draws it. There must be a drawing power in matter: and the sum of the drawing power in the matter of the earth must be in the earth's centre, not in any side of the earth. Therefore does this apple fall perpendicularly, or towards the centre. If matter thus draws matter, it must be in proportion of its quantity. Therefore the apple draws the earth, as well as the earth draws the apple. That there is a power, like that we here call gravity, which extends itself through the universe.[7] It was another twenty-two years before that 'inspiration' was fulfilled with the publication of the *Philosophiae Naturalis Principia Mathematica* in 1687.

The image of the dreamer appears to have appealed to the ageing Newton for he perceived his life's work as a very romantic discovery. He claimed

> I do not know what I may appear to the world; but to myself I seem to have been only like a boy playing on the sea-shore, and diverting myself in now and then finding a smoother pebble or a prettier shell than ordinary, while the great ocean of truth lay all undiscovered before me.[8]

This image projected Newton not only as a romantic dreamer, but also as tolerant, patient and devoted to pets. There is a story of a fire in Newton's laboratory that took place sometime between 1677 and 1683. According to the story, Newton's dog, Diamond, knocked over a candlestick, which had resulted in the fire. Newton lost many manuscripts including his work on the *Philosophiae Naturalis Principia Mathematica* and *Opticks*, and he is said to have "worked them over again".[9] Despite this devastating loss, Diamond was only rebuked by Newton, who exclaimed "Oh Diamond! Diamond! Thou little knowest the mischief thou hast done".[10] However, according to Newton's amanuensis Humphrey Newton (no relation), Newton never had any pets.[11] Although in his scientific work, Newton did suggest that diamonds might be made to burn, because of their unique optical properties[12] (perhaps indicating that the myths of Newton stemmed from his generally misunderstood scientific work). Despite the lack of truth in having Diamond the dog, the myth of the romantic dreamer image is still perpetuated to this day and a favourite question at trivia night quizzes is "What is the name of Isaac Newton's dog".

The nineteenth century images of Newton show him lost in his thoughts. The epitome of this image is *Isaac Newton at the age of Twelve* by Frederick Newenham.[13] The 12-year-old Newton is pensively gazing out into space reflecting on the man he will became, this engraving was executed with all the splendour of Victorian romanticism. Equally important are the depictions of Newton's religious devotion. In 1820, an English painter Benjamin Robert Haydon unveiled what is considered to be his masterpiece, *Christ's Entry into Jerusalem*; Newton is in the crowd, along with other notable figures such as John Keats and William Wordsworth, admiring the figure of Christ.[14]

But throughout his life Newton was not this figure of a romantic dreamer. The depth of his life's work is revealed by his two great scientific texts: *Philosophiae*

Naturalis Principia Mathematica, generally truncated to *Principia*, **which was** first published in 1687 (with a further two revised editions in his lifetime in 1713 and 1726) and *Opticks,* first published in 1704. In addition, there were hundreds of unpublished manuscripts. The depth and scope of his life's work is astounding, and reveals him to be a true polymath and genius.

The Unpublished Manuscripts of Isaac Newton

Newton died on 20th March, 1727 leaving hundreds of unpublished manuscripts; some of which date back to his arrival at Trinity College, Cambridge in 1661. His heirs invited Thomas Pellett to examine the manuscripts and report on their suitability for publication. After just three days of examining these hundreds of manuscripts, Pellett, a qualified physician and member of the Royal Society, dismissed the majority of the manuscripts as being "not fit to be printed",[15] "of no scientific value" and "loose and foul papers".[16]

Pellett only found two sets of manuscripts suitable for publication. The first was a set of manuscripts on chronology and the second were two manuscripts on prophecies. Although Pellet claimed that the text on prophecy was imperfect, they were nevertheless worthy of publication. The other manuscripts, which included drafts of the *Principia*, mathematical and scientific papers, his correspondence and works on prophecy, chronology, alchemy and theology were passed on to his niece Catherine Conduitt. With the marriage of Catherine's daughter into the Portsmouth family, the manuscripts become part of the Portsmouth Collection.

In 1872, the papers were offered to the University of Cambridge, which only accepted the scientific papers, refusing the other papers on topics that Newton was not famous for.[17] The remaining non-scientific manuscripts were offered to the British Library, which also refused them on similar grounds. These manuscripts remained in the Portsmouth Collection until 1936, when they were auctioned and dispersed into collections all around the world.

The auction was held in July in 1936 at Sotheby's. The manuscripts were divided up into three-hundred and thirty lots and sold to thirty-three buyers.[18] Thus Newton's manuscripts were scattered all over the world. It is surprising that these manuscripts were allowed to leave England. Josè Faur considered that the reason for this was because of the contents of the manuscripts. Manuscripts on prophecy, alchemy and Newton's unorthodox theology did shock some scholars. It was "to protect Newton's 'good name,' [that] the importance of the manuscripts were denied".[19]

One of the buyers was the famous economist John Maynard Keynes, who bought a significant number of manuscripts which he bequeathed to King's College, Cambridge. He made a study of these manuscripts and found

> that Newton was different from the conventional picture of him. But I do not believe he was less great. He was less ordinary, more extraordinary than the nineteenth century cared to make him out. Geniuses *are* very peculiar.[20]

The nineteenth century, in their adulation of Newton, had rendered him quite bland. After poring over the contents of the box of manuscripts he had purchased, Keynes claimed:

> Newton was not the first of the age of reason. He was the last of the magicians, the last of the Babylonians and Sumerians, the last great mind which looked out on the visible and intellectual world with the same eyes as those who began to begin to build our intellectual inheritance rather less than 10,000 years ago.[21]

This very famous quote was written by Keynes in 1942. The paper "Newton the Man" was written for the tercentenary celebration of Newton's birth, but the Second World War intervened and the paper was not presented until 17 July 1946, after Keynes' death in April of that year. These tercentenary celebrations were conducted on an international scale and ran for five days culminating in a garden party at Buckingham Palace.[22] Keynes' paper was read to the Royal Society by his brother, Geoffrey. It had not been revised by the author, who had written it some years back.[23] Keynes was the first to publicly consider Newton as more than the orthodox image of the romantic dreamer scientist. He considered Newton's faults and also what appeared from an early twentieth century perspective to be unorthodox practices such as alchemy, his style of theology, his interest in chronology and church history and his argumentative nature, in conjunction with his great scientific achievements. These were aspects of Newton's character that had been ignored or glossed over by previous commentaries and biographies.

These discoveries in *Newton the Man* by Keynes did not lessen his admiration for Newton. He considered that the box of papers that he was studying showed Newton to be a man with great power of mind, who attempted to understand all aspects of God and nature. Newton's experiments were not undertaken for mere discovery, but to verify what he already knew, and to confirm his strong belief in God. Keynes wrote:

> Why do I call him a magician? Because he looked on the whole universe and all that is in it *as a riddle*, as a secret which could be read by applying pure thought to certain evidence, certain mystic clues which God had laid about the world to allow a sort of philosopher's treasure hunt to the esoteric brotherhood. He believed that these clues were to be found partly in the evidence of the heavens and in the constitution of elements (and that is what gives the false suggestion of his being an experimental natural philosopher), but also partly in certain papers and traditions handed down by the brethren in an unbroken chain back to the original cryptic revelation in Babylonia. He regarded the universe as a cryptogram set by the Almighty – just as he himself wrapt the discovery of the calculus in a cryptogram when he communicated with Leibnitz. By pure thought, by concentration of mind, the riddle, he believed, would be revealed to the initiate.[24]

As more and more of Newton's papers became available to scholars, Keynes' words seem increasingly insightful and revealing. Keynes considered that there were two sides to Newton's character they were "Copernicus and Faustus in one"[25]. Scientist and magician were the same man working to one purpose and whose achievements were seemingly beyond his era but at the same time founded in the knowledge of the ancients.

Later biographies have assumed that the works on theology, chronology and prophecy were the works of an ageing Newton. That there were two Newtons; the great scientist of his youth and the ageing Newton who had lost his taste and ability for science and turned to the study of chronology, prophecy and religion as a result of the nervous breakdown he suffered in 1693.[26] However, these two separate and diverse personas are not supported or divided by any such date and Newton did continue to research and add to the science of his day. Furthermore, his papers and interest in chronology and prophecy date back to his earliest days in Cambridge in the 1660s.

Newton's deeply held religious convictions led him to search for the mystic clues which he believed that God had laid about the world. This search had resulted in his scientific research in the form of the *Principia* and *Opticks;* both are landmarks in science. Alchemy, chronology, theology and prophecy as well as natural philosophy were all parts of these clues which Newton attempted to unravel or decrypt. It is unclear whether there was a dividing line in the mind of Newton between these topics; however, all of these topics confirmed his belief in the supreme design of the universe.

Argument from Design

In the *Principia,* Newton had established the foundations of classical mechanics and the law of universal gravitation that enforced the notion of a clockwork universe. But for Newton, these laws of physics revealed that the universe was designed by a Supreme Being. He stated in the *Principia*

> When I wrote my treatise about our System I had an eye upon such principles as might work with considering men for the belief of a Deity & nothing can rejoice me more than to find it useful for that purpose.[27]

This designed universe could not arise from "purely mechanical" causes. In the "General Scholium" of the *Principia,* Newton had a Biblical interpretation of God's role in universal physics. He asserted that

> This Being governs all things, not as the soul of the world, but as Lord over all; and on account of the dominion is wont to be called *Lord God*, or *Universal Rules*. . . The Supreme God is a Being eternal, infinite, absolutely perfect; but a being, however perfect, without dominion, cannot be said to be Lord God; for we say, my God, your God, the God of Israel, the God of Gods and Lord of Lords, but we do not say, my Eternal, your Eternal, the Eternal of Israel, the Eternal of Gods. . .. And from his dominion it follows that the true God is a living, intelligent, and powerful Being; and, from his other perfections, that he is supreme, or most perfect. He is eternal and infinite, omnipotent and omniscient; that is, his duration reaches from eternity to eternity; his presence from infinity to infinity; he governs all things, and knows all things that are or can be done. He is not eternity and infinity, but eternal and infinite, he is not duration or space, but he endures and is present. He endures forever, and is everywhere present; and, by existing always and everywhere, he constitutes duration and space . . .. It is allowed by all that the Supreme God exists necessarily; and by the same necessity he exists always and everywhere. . . We know him only by his wise and excellent continuance of things and final causes; we admire for his perfections; but we reverence and adore him on account of his dominion: for we adore him as his servants; and a god without dominion, providence, and final causes, is nothing else but Fate and Nature. . . Since every

> particle of space is *always*, and every indivisible moment of duration is *everywhere*, certainly the Maker and Lord of all things cannot be *never* and *nowhere*.[28]

In the "General Scholium", Newton was deriving God neither from personal experience nor from reason, but when he stated that God was dominative, omnipresent, transcendent and beyond all human reason, Newton was expressing a traditional Jewish view of God, and this possibly reveals the influence of the great medieval Jewish theologian Moses Maimonides on Newton.[29]

Newton's unpublished manuscripts reveal that he did not believe in the Trinity; he believed that Christ is the first amongst creatures and he was the mediator between God and man. He was a prophet; and essentially Christ the Son was not the same essence as the God the Father.[30] To prove his religious convictions, he studied chronology and prophecy in search of the original religion. Newton perceived that Christianity had become corrupt through time and that he lived in idolatrous times.[31] Keynes claimed that

> Very early in life Newton abandoned orthodox belief in the Trinity. . . . It may be that Newton fell under Socinian[32] influenced circles, but I think not. He was rather a Judaic monotheist of the school of Maimonides. He arrived at this conclusion, not on so-to-speak rational or skeptical grounds, but entirely on the interpretation of ancient authority. He was persuaded that the revealed documents give no support to the Trinitarian doctrines which were due to late falsifications. The revealed God was one God.[33]

The main falsification being the "deliberate corruption of the New Testament texts by wicked characters like Saint Athanasius",[34] who argued against Arius' doctrine that Christ was distinct from God the Father, at the first Council of Nicaea in AD 325.

Newton studied a wide range of Jewish texts by Josephus, Philo, Maimonides and the Talmudic scholars and took an interest in Jewish ritual observances, the Temple of Jerusalem, the size of the cubit and Biblical prophesy. However, this interest was not confined to Newton. In the humanist movement of the Renaissance and Early Modern era, there was a rise of interest and scholarship in Hebrew and Jewish texts. Many scholars such as Giovanni Pico Della Mirandola, Giordano Bruno, Marsilio Ficino and Gerardus Vossius thought that the ancients held the secrets of the world as well as the solutions to the problems that plagued Christianity. Newton did not consider all Jewish texts to have work of merit and some he considered heretical, the Cabbalists in particular. Newton considered that heresy of the pure religion was widespread in the Middle East and spread by the heathen Greek philosophers and the Cabbalist Jews.[35] His studies of the Jewish texts were restricted to the historic strands of Judaic Biblical interpretation and to what Newton perceived to be a more rational approach of philosophers such as Maimonides and his commentators such as Gerardus Vossius.[36]

In the "General Scholium", Newton stressed the supreme dominion of God. Comprehensive studies have been made on the similarities between Newton and Maimonides as an explicit influence.[37] Maimonides

> promoted the idea that there is total harmony between Scripture and natural science. Each, he believed, has two distinct levels of study and comprehension: an exoteric level accessible to all, and an esoteric level accessible only to the elite.[38]

Newton applied this concept to his studies of physics and his theological work.

For Maimonides the esoteric elements of the physical sciences were encoded by the prophets in metaphors and riddles. Although they were accessible to all, they could only be apprehended by the wise. To reveal these secrets "to the unlearned would be as harmful as feeding a baby 'wheat-bread, meat, and wine' ".[39] Newton's interest in an original and universal religion led him to the Jewish doctrine of God's supreme dominion and to an interest in Maimonides codification of the Jewish rituals and artefacts of the temple. Although some influence by Maimonides is evident, Newton did not holistically follow the Maimonidean school of thought as was suggested by Keynes; particularly Newton's plan and metaphoric concept of the Temple of Jerusalem, which is distinctly different from Maimonides'.[40]

Newton was interested in the Hebrew language; he owned Hebrew dictionaries and lexicons as well as several books and Bibles in Hebrew.[41] However, just how good his Hebrew was is disputed.[42] In Babson MS 0434, Newton did not quote Hebrew, he only used single words and the Hebrew roots that he referred to are in the margins of the *Biblia Sacra Polyglotta* which was in his library.[43] Notably, most of Newton's books on the Jewish religion and history[44] are translations, including five of Maimonides' works, which were also in his library.[45] He made numerous references to these five books and all of these were Latin translations, so that he would have been familiar with Maimonides' works without the need of a fluent Hebrew or for that matter without the need of any Hebrew. In Newton's library there was Vossius' three-volume commentary *De Theologia Gentili* which was a translation of Maimonides' *Misheh Torah, Hilk Avodah Zarah* written by Vossius' son Isaac. The pages of this are described as being "very extensively dog-eared with 112 pages still turned down and several similar signs".[46]

As for the vision of the prophets, Newton chose

> those constructions which without straining reduce things to the greatest simplicity. The reason of this is manifest by the precedent Rule. Truth is ever to be found in simplicity, & not in the multiplicity & confusion of things. As the world, which to the naked eye exhibits the greatest variety of objects, appears very simple in its internal constitution when surveyed by a philosophic understanding, & so much the simpler by how much the better it is understood, so it is in these visions. It is the perfection of God's works that they are all done with the greatest simplicity. He is the God of order & not of confusion. And therefore as they that would understand the frame of the world must endeavour to reduce their knowledge to all possible simplicity, so it must be in seeking to understand these visions.[47]

Order and simplicity were keys to understanding the metaphors and riddles of the prophets and to the frame or structure of the world.

In the *Principia,* Newton used rules of reasoning to build up propositions, lemmas and corollaries showing these could be demonstrated as pure mathematics structures – as geometrical constructs. In other words, he built up a lexicon – a new mathematics to be able to understand natural philosophy. As an undergraduate at Cambridge, Newton became acquainted with Robert Sanderson's *Logicae artis compendium.*[48] Newton used Sanderson's model of logic in both his natural philosophy and his scriptural exegesis of the Apocalypse. Maurizio Mamiani claimed that

> The methodological affinity between scientific and religious knowledge is so narrow in Newton's works that it seems highly improbable that chance is at work. Both science and religion have a common root and a common reference to certainty.[49]

There is a strong correspondence between the rules of reasoning in the final version of the *Principia* of 1726 and the sixteen rules for the interpretation of the Scripture he compiled around the 1670s.[50] Harmony between the Scriptures and the analogies of the prophets are equivalent to the simplicity of nature.[51] In an unpublished projected preface to the *Principia,* Newton linked his methodology with that of the ancients. He claimed that the ancients had two methods of mathematics which

> they called synthesis & analysis, or composition & resolution. By the method of analysis they found their inventions & by the method of synthesis they (published them) composed them for the public... The propositions in the following book were invented by analysis. But considering that (they were) the ancients admitted nothing into geometry (but that) before it was demonstrated by composition I composed what I invented by analysis to make it (more) geometrically authentic & fit for the public.[52]

There is a similarity to Sanderson's *Logicae artis compendium* that made the distinction between method of discovery (analysis) and method of presentation (synthesis).[53] Knowledge begins with analysis then progresses to synthesis. The analysis, which consisted of empirical experiments and observation, is carried out according to strict rules and derived experimental principles. The synthesis is a deduction of future occurrence based on these principles. This procedure is revealed in Query 31 in *Optick* published in 1704:

> As in mathematics, so in natural philosophy, the investigation of difficult things by the method of analysis, ought ever to precede the method of composition. This analysis consists in making experiments and Observations, and in drawing general conclusions from them by induction, and admitting of no objections against the conclusions; yet it is the best way of arguing which the nature of things admit of, and may be looked upon as so much the stronger, by how much the induction is more general. And if no exception occurs from phenomena, the conclusion may be pronounced generally. But if at any time afterwards any exception shall occur from experiments, it may then begin to [be] pronounced with such exceptions as occur. By this way of analysis we may proceed from compounds to ingredients, and from particular causes to more general ones, till the argument end in the most general. This is the method of analysis: And the synthesis consists in assuming the causes discovered and established as principles, and by them explaining the phenomena proceeding from them, and proving explanations.[54]

In the "General Scholium", Newton stated "Supreme God exists necessarily; and by the same necessity he exists always and everywhere". For Newton, God is the "First Cause," the "First Principle" and all of creation was subordinate to God.

For Newton, neither would the universe arise from "purely mechanical" causes nor would God design a mechanical universe. It would not just be wound up and left to run. As proof, Newton noted small variations in the orbits of Jupiter and Saturn.[55] Also in *Opticks,* Newton claimed that comets could not keep their orbs.[56] Newton concluded that the designer of the universe, God, had to intervene occasionally to "repair" and "restore" the balance of the universe. God was not only the first cause, but He continually sustained the universe – God was a necessary present.

In the *Principia,* Newton used rules of logic and complex mathematics to reveal the geometric simplicity of the structure of the world. This took the form of demonstrating the geometrical underpinning of the conic as the path necessary for establishing both the orbit of planets and of comets – for Newton it was God's plan and domain, and it is only though God's intervention that the universe kept going. Newton's theology is intertwined with his natural philosophy and it is problematic to attempt to separate them. Newton's sixteen rules to interpret the prophets predate the writing of the *Principia* and they have a similar methodology to the rule of reason that is found in the *Principia.* This is not to say that Newton consciously analysed natural philosophy in terms of theology but that the interaction between them "is a two-way street".[57] While theology and natural philosophy was a two-way street for Newton, for Andrew Cunningham the distinction between God and natural philosophy is erroneous.[58] Natural philosophy should not be considered as science in the modern sense. He claimed that "natural philosophy as such was a discipline and subject-area whose role and point was the study of God's creation and God's attributes".[59] In Query 31 in *Opticks,* Newton claimed that

> if natural Philosophy in all its parts, by pursuing this method (of analysis and composition), shall at length be perfected, the bounds of moral philosophy will be also enlarged. For so far as we known by natural philosophy what is the first cause, what power he has over us, and what benefits we receive from him, so far our duty towards him, as well as that towards one another, will appear to us by the light of nature.[60]

This was written in 1704 and the General Scholium nearly a decade later; however, the argument from design was established very early in Newton's scientific unpublished papers.

The "General Scholium" was not added until the second edition of the *Principia* in 1713; however, Newton's rationale behind "General Scholium" can be seen in a much earlier work from his unpublished papers. *De Gravitatione et Aequipondio Fluidorum*[61] has been dated as early as 1664 but certainly before 1673.[62] In *De Gravitatione,* Newton described a Divine cosmogony where bodies exist because of Divine Will. "Space is eternal in duration and immutable in nature, and this is because it is the eminent effect of an eternal and immutable being".[63] The movement of bodies in space was "nothing more than the produce of the Divine Mind realized in a define quantity of space".[64] God informs bodies which are inherent in space and they move according to certain rules.

These concepts were established very early in Newton's intellectual life and were developed in the most productive time of his intellectual life. In the period from the mid-1660s to the mid-1680s, Newton produced works on theology, prophecy, the history and corruption of the Church, chronology and Solomon's Temple prior to and at the same time that he was working on the *Principia.* The concepts behind these works stayed with him until his death in 1727. The later drafts of his unpublished papers were rewritings of the earlier original works and in general these later drafts were "sanitised" and they were made more conservative by Newton himself presumably for public consumption.

Structure of This Book

The aim of this book is to put Newton's works on the Temple of Solomon into perspective in relation to his other works. From his unpublished manuscripts, it appears that Newton's interest in the Temple of Solomon began at least in the early 1680s and continued throughout his life. Newton considered it important enough to attempt to reconstruct the Temple from Biblical and historic sources, and at the same time that he was writing the *Principia.* He was also spending a considerable amount of time researching the design of the Temple, the length of the Hebrew cubit and the rituals of the Temple. His unpublished papers reveal a large volume of work on or related to the Temple. Yet his work on the Temple has been largely ignored; the purpose of this book is to redress the balance and to demonstrate the significance of Newton's research of The Temple in the context of Newton's overall contribution to science.

Chapter 2, "Chronology, *Prisca Sapientia* and the Temple" reviews Newton's interest in chronology and the controversy that arose at the end of Newton's life over his chronology. The purpose of this chapter is to place Babson MS 0434 into context with his other writings and the writing of contemporaneous writers on chronology and *Prisca Sapientia*. After his death, the executors of his will published two books from his unpublished papers; one on chronology and the other on prophesy. *Chronology of Ancient Kingdoms Amended* was compiled from at least sixteen drafts left by Newton. This chapter examines how he considered the actions and events of the Egyptians and the ancient Greek heroes to be measured and dated by the death of a king of another nation, King Solomon. It also considers the importance of *Prisca Sapientia*, or ancient wisdom, to Newton.

The seventeenth century had a rich tradition of prophetic and apocalyptic writings. Chapter 3 "Prophesy and the Temple" examines the influences on Newton's work on prophesy. Newton built on the apocalyptic writings of Mede and More. He attempted to decode the hieroglyphic language of the prophets that was encoded in the Scriptures. In this hieroglyphic language, there is a correspondence between the symbols of heaven and earth. The earthly rituals and parts of the Temple correspond with the celestial city of the future and the Temple was the scene or background for the visions of the prophets. This chapter then examines Newton's unpublished work on the Prytanæa, an ancient religion that was closely connected to the original religion of Noah and retained the knowledge of natural philosophy in their rituals. To Newton the practices and the rituals of the Prytanæa represented the frame of the world and they retained the true knowledge of natural philosophy. Finally, the question, "in the seventeenth century was Solomon's Temple the same as Ezekiel's vision of the Temple" is examined.

Chapter 4 "Prytanæum and the Floor Plan of the Temple of Solomon" further expands the theme of Newton's concepts of the Prytanæa's rituals. These expressed the lost *Prisca Sapientia* of natural philosophy which stemmed from the original religion of Noah. The frame of the Prytanæum or Temple embodied the original religion and symbolised the geometric structure of the universe. This geometric structure was the mathematical form of the universe untainted by the corruption of the original religion. In turn, the Temple of Solomon, which replicated the plan of the Tabernacle of Moses, embodied the perfection of the original religion within its structure, which had been inherited from the time of Noah. To understand the frame of the Temple was to understand a great part of the original religion's theology; the frame of the Temple was the symbol of the exoteric knowledge; while the enacting and understanding of the rituals within the Temple led to the esoteric knowledge of the prophets and represented the structure of the heliocentric solar system.

This chapter then considers the seventeenth century's concepts of Solomon's Temple as a microcosm of the macrocosm and Newton's commentary on Juan Battista Villalpando. Newton mentioned Stonehenge as an example of an English Prytanæum; some examples of this concept are examined; in particular, Inigo Jones's reconstruction of Stonehenge. Finally, the differences between the great medieval Jewish theologian, Moses Maimonides and Newton's floor plans of the Temple are considered.

Chapter 5 "The Temple Measurements and the Sacred Cubit" analyses an important but neglected work of Newton on the Hebrew sacred cubit, "A Dissertation upon the Sacred Cubit of the Jews and the Cubits of the Several Nations". "Dissertation" was published in 1737 ten years after Newton's death. It is a working paper and was never intended for publication in its original form. Although the paper is flawed, it is ingenious. Despite its flaws, "Dissertation" is a highly significant manuscript. It gives an insight into Newton's vast understanding of ancient sources and measurements, and his working methods.

Chapter 6 "An Overview of the Contents and the Source of Babson MS 0434" outlines the structure of the manuscript, then it examines the sources of Babson MS 0434, which further reveal the depth of Newton's knowledge and reading. Like "Dissertation", Babson MS 0434, is a working document and it was never completed.

Chapter 7, "Reconstruction", considers the illustrated floor plans of the Temple in the *Chronology* and Babson MS 0434. Next an examination of the differences in the descriptions of the Temple in the *Chronology* and Babson MS 0434 is undertaken. These differences call into question whether the chapter of the Temple in the *Chronology* is the work of Newton. Then the developments of the Temple plan in Babson MS 0434 are considered. Finally, a reconstruction of the two- and three-dimensional plan, as described by Newton in Babson MS 0434, is presented. The commentary concludes by considering the role of the Temple within the work of Newton.

Finally, the translation of Babson MS 0434 or "A Treatise or Remarks on Solomon's Temple Introduction to the Lexicon of the Prophets, Part two: About the appearance of the Jewish Temple" is presented along with the parallel pages of the Latin text. In this Latin text, Newton's erratic spellings and his equally variable punctuations have not been corrected.

Conclusion

Newton, the natural philosopher, is often separated from Newton the theologian, the alchemist and the chronologist even to this day. His works *Principia* and *Opticks* are not separate entities from his other works and if there is a dividing line between these two personas, it is nearly invisible. Newton's unpublished manuscripts revealed a far more interesting and complex man than the bland genius dreamer that history has portrayed him to be.

He regarded the universe as a cryptogram that was set by the Almighty and he attempted to decode the puzzle. However, that decoding was on two levels, the esoteric and the exoteric. Both levels were equally important to Newton and both were indivisible. For above all, Newton searched for the truth and that truth was the one plan of God in which everything was linked. Newton used a similar model of logic for his natural philosophy and his scriptural exegesis. The Temple was an integral part of the puzzle, and his studies of it were wide ranging. Babson MS 0434 concentrates on the architecture of the Temple but it is not an isolated manuscript on the Temple, and it relates to many other works of Newton.

Newton understood his work in the *Principia* to be part of God's grand design. However, he did perceive that this knowledge of the physical universe was exoteric knowledge and was accessible and apprehensible to all. But God encoded the spiritual universe with analogies and cryptograms that reveal His will. Although these were accessible to all through the words of the prophets, a lexicon that was used by all the prophets, such esoteric knowledge could only be decoded by the wise. Newton told John Conduitt that he believed that the study and understanding of the original religion would resolve the religious disputes of theology in the same way as the *Principia* had resolved the questions of natural philosophy.[65] For Newton, the Temple of Solomon played an important role as an analogy of God's grand design, and in the decoding of the cryptograms of the Temple it gave insight into his design. "For there is no way (without revelation) to came to the knowledge of a Deity but by the frame of nature".[66]

Chapter 2
Chronology, *Prisca Sapientia* and the Temple

Apart from *Reports as Master of the Mint*, which were published between 1701 and 1725, Newton published only scientific manuscripts in his lifetime. *Principia* published in 1687, which consisted of three volumes, defined the three laws of motion and gravitation which lay down the foundation of classical mechanics, and in the formation of this theory he developed the mathematical field of calculus. Newton added material and revised the *Principia* in 1713 and 1726. His second contribution to science was *Opticks*, which considered the properties and the refraction of light, and was published in 1704. These two books had established Newton as the most significant scientist of his time. However, science was not his only interest and in fact Newton's library consisted of only 52 volumes, or 3% of the whole library, on mathematics, physics and optics.[67]

At the time of Newton's death, his library consisted of approximately 2100 volumes of which 1763 had been accounted for in a study made by John Harrison published 1978.[68] Newton did not make inventories of his library as was the fashion of the day by professional book collectors[69]; his books were working books – tools. They were often heavily annotated and showed excessive use. The most heavily annotated book in his library was his English Bible which had been bound together by the Book of Common Prayer and a version of Psalms. The Books of Daniel and Revelations showed the most frequent use and the most marginalia.[70] William Stukeley in his reminiscences of Newton claimed that "No man in England read the Bible more carefully that he did, none study'd it more…the Bible which he commonly used, thumbed over, as they call it, in an extraordinary degree, with frequency of use".[71] There are 477 volumes in his library, or 27.5% of the whole library, on theology. This is more than any other subject.[72] The rest of the library was a mixture of chronology, history, classical history and commentaries, philosophy, alchemy, twenty-four volumes that dealt directly with Judaism, and many others that contained Jewish history.[73] Two of the earliest purchases that Newton made on arriving at Cambridge University in 1661 were Hall's *Chronicles* and Johann Sleidan's *Four Monarchies,*[74] and the Sleidan remained in his library for the rest of his life.[75] Chronology, particularly associated with prophecies, remained a life-long interest.

Robert Hooke had written to Newton in Cambridge in November of 1679 asking Newton's opinion on his and others' theories of celestial motions of the planets.[76]

T. Morrison, *Isaac Newton's Temple of Solomon and his Reconstruction of Sacred Architecture*, DOI 10.1007/978-3-0348-0046-4_2, © Springer Basel AG 2011

Newton had been in Lincolnshire attending to a family matter after the death of his mother and he replied to Hooke on his return to Cambridge. He claimed that

> I have had no time to entertain Philosophical meditations or so much as to study or mind any thing else but country affairs. And before that, I had for some years past been endeavouring to bend my self from Philosophy to other studies in so much that I have long grouched the time spent in that study unless it be perhaps at idle hours sometimes for a diversion.[77]

His interest in *Prisca Sapientia*, or ancient wisdom, is reflected in his library. His studies on alchemy and in his unpublished papers reveal a reverence for the ancient wisdom of the *veteres* (ancients). The existence of Newton's writing on theology, ancient philosophy and mythology was known by his Scottish friend David Gregory and Archibald Pitcaire in at least 1694–95.[78] Newton applied his knowledge to consider the chronology of ancient civilizations and in particular the development of these civilizations. Chronology was a subject that was an established scholarly genre.

Kings had genealogies drawn up that linked them to heroic and mythical characters such as Aeneas of Troy to establish not only their claim to the throne, but also their impressive and divine linage.[79] Chronology could be used to prove a nations antiquity.[80] In 1689, Olof Rudbeck's *Atlantica* was published. In it he claimed that his home country Sweden was Atlantis and thus was the cradle of civilization.[81] In the seventeenth century, there was a strong interest in the chronology of world history. Sir Walter Raleigh's *History of the World,* published in 1614, was a massive work in two volumes which remained popular throughout the seventeenth century and was reprinted in its original and abridged form eight times in the next hundred years. James Ussher's *The annals of the World*, published in 1658, was a highly significant book that considered the origins, along with the sacred and profane history, of the ancient world. Newton worked on the chronology of the ancient world and he attempted to put it on a scientific footing. At the time his interest in chronology was known, but it was considered by many that his output on this subject was minimal; that it was a very secondary study compared with his study of science. After all, he was the man who understood and revealed the mechanics of the universe. In fact, his unpublished manuscripts showed that there were close to 200,000 words on chronology[82] and he continued to work on the chronology of the ancient world until the end of his life. It also drew him into controversy which continued beyond his life time.

The Short Chronicle

In 1716 Princess Caroline, an admirer of Newton, persuaded him to give her a copy of his chronology. Although he delayed giving it to her with excuses that it was "imperfect and confused" he had little choice but to comply with the royal command.[83] He drew up an "Abstract" of his chronology, which was barely twenty pages long and was primarily comprised of a list of names and dates with no justification of how those dates were worked out. There had been other interests in obtaining a copy of the "Abstract" from Abate Conte Antonio Conti, a Viennese nobleman commonly known as Abbé Conti, but Newton refused to provide a copy.[84] Abbé Conti approached Princess Caroline who requested that Newton should supply him with a copy. Again Newton complied with the royal command. Newton did, however, request that Abbé Conti keep the work private.[85] At this stage, Newton must have thought that that was the end of the "Abstract".

Despite promising to keep the "Abstract" a secret, once Abbé Conti had left England, a year later, he broke this promise. He showed it around to the "learned circles of Paris". Amongst the learned authorities of ancient chronology that he showed it to was the Jesuit Priest Etienne Souciet. Souciet had many queries about the Newton chronology and these were shown to Newton by an Oxford astronomer, John Keill, in 1720,[86] so Newton was aware the copies of the "Abstract" were being circulated. Furthermore, there is evidence that Newton himself gave copies to others who, in turn, generated more copies.[87] There are three surviving handwritten copies in Cambridge.

By 1724 the chronology passed into the hands of Nicolas Fréret, an eminent scholar of antiquities at the Académie des Inscriptions et Belles-Lettres in Paris. He was highly critical of Newton's scheme, claiming it to be too limited. Fréret translated it into French and he wrote a critical commentary entitled *Abrégé de la chronologie* (Summary of the Chronology). Then he began to arrange to have it published.[88]

There had been little in the way of archaeological evidence to support any chronology of the ancient civilisations in the early eighteenth century. The spectacular finds unearthed at Herculaneum did not begin until 1749[89] and the Egyptian hieroglyphs were not translated until Jean-Francois Champollion deciphered them in 1824.[90] Prior to the early eighteenth century, the most significant find, as far as chronology was concerned, was the Parian Marble. The Parian Marble was acquired by the Earl of Arundel and shipped to England in 1627. It listed the dates for a number of kings and chief magistrates of ancient Athens from 1582 BC to 264 BC. The Parian Marble dated major events such as the invention of corn by Demeter to 1409/8 and the fall of Troy to 1209/8.[91] Chronologist believed it to be the missing link that would make sensible the other material that they had collected.[92] Otherwise, most chronologies were composed with the aid of literary texts. These included the work by Herodotus, Apollodor, Theodotus, Josephus, Diodorus, Strabo, Plutarch and the Bible.

In the early chronologies, myths were reality and marked major events in history. In Newton's "Abstract" he listed the dates of key events in history such as:

> 989 (BC) Dædalas and his nephew Talus invent the saw, the turning-lath.
> 988 Minos makes war upon the Athenians for killing his son Androgeus.
> 987 Dædalas kills his nephew Talus, and flies to Minos.[93]

Although Newton did not mention the Minotaur the existence of the mythical characters of Dædalas, Talus, Minos and many others throughout his chronology, and the surrounding events, were not in question. He believed that the figures of mythology were in fact "heroes" of past history. Although their actions had been embellished in the telling, over time they had existed and there was some truth to their actions. The dates of these heroes marked important turns in history; particularly, technological developments such as the invention of the saw and the turning lath. Only a few criticized Newton for this use of mythology in his chronology, one being William Whiston, former pupil and successor to Newton as Lucasian Professor at Cambridge. He claimed "that tho' it be a work of vast learning and very uncommon sagacity, yet is it built, not upon ancient evidence, and the testimonies of historical authors; but partly upon the poetic stories of Mythologists".[94] After the English publication of his chronology there was a comparison made in a contemporary journal of the chronology of Bishop Cumberland and that of Newton.[95] Much of the discussion generated was not over the fact that Newton and Cumberland had both based their arguments on mythology, but rather which one of them presented the correct chronology of mythology.

The major criticism of Newton's system was the dating of these events, and not the inclusion of the myths. Newton had dated the Argonautic expedition to 936 BC, 43 years after the death of Solomon,[96] and the fall of Troy to 76 years after the death of Solomon.[97] He shortened the accepted Greek history by at least 300 years. He had also shortened the duration of the Egyptian Empire and claimed that they had overstated their antiquity beyond the age of the world.[98]

Fréret's French translation of the "Abstract" and his commentary came into the hands of a Parisian Printer Guillaume Cavelier who immediately wrote to Newton, requesting any remarks he had on Fréret's translation and commentary, or corrections or additions, before he published it.[99] He wrote twice without any reply from Newton, the third time he stated that since he had written twice previously that if he did not hear from him he would take Newton's silence as confirmation of his consent.[100] Although Newton did eventually write withholding his permission to publish, Cavelier claimed that he received Newton's letter too late to stop the publication of it. It was published in 1725 as the *Abregé de la Chronologie* and Newton received a copy from Cavelier on 11th November 1725.[101]

Newton reacted by writing "Remarks on the Observations made on a Chronological Index of Sir Isaac Newton"; this was translated into French by the Observer and published in Paris (with exceptional speed) in December 1725 in *Philosophical Transactions of the Royal Society.*[102] In "Remarks" he stressed that there was no permission from him to print the work and observed:

> as if any man could be so foolish as to consent to the publishing of an unseen translation of his papers, made by an unknown person, with a confutation annexed, and unanswered at the first appearance in public.[103]

This was followed by a refutation of several of the observations which had been added to the French translation by Fréret. He primarily blamed Abbé Conti for the entire situation. The "Remarks" was ill-considered and it displayed all the malice that he revealed in his treatment of John Flamsteed, the first Astronomer Royal of England (Clark and Clark, 2000)[104] and in the debate over calculus with Leibniz.[105] His "Remarks" also caused as much controversy as the *Abregé de la Chronologie.*

In the controversy with Gottfried Wilhelm Leibniz who had invented calculus, there was a clear division of opinion between the continent and England. With this new controversy, the division was not so clear. Newton's dates contradicted all of the established chronological works including the work of Bishop Ussher. Ussher had established the date of creation to be 6.00 PM 22nd October, 4004 BC.[106] In 1701, Ussher's chronology was given the blessing of the Church of England[107] and this date was printed into the Bible as the date of creation until the turn of the twentieth century.[108] But it was not the date of creation that Newton contradicted; it was in the dating of the historic events that Newton moved away from accepted convention. Ussher had established the date of the Trojan War to be 1184 BC as compared with Newton's date of 903 BC. Even Newton's closest allies had difficultly supporting his dates. Stukeley claimed that Newton's "chronology was somewhat very particular, and likewise solid. But whilst he has justly shortened the years of the world, he appears to me to have done it a little too much".[109] He was more critical in his personal letters. In a letter to Antiquarian Roger Gales dated 1728 he stated "Mr Conduitt has sent me Sir Isaac Newton's Chronology: I don't admire his contracting the spaces of time; he has pursued that fancy too far".[110]

As the controversy began to rage, more commentaries came out against Newton's "Abstract", which had been re-titled "A Short Chronicle". Father Souciet joined in by claiming that "M Newton's Chronology cannot stand. . .he has made an error of about 530 years. . .mine on the contrary is correct".[111] The amount of years that Newton had shortened the chronology was in dispute, but generally it was accepted that he had shortened it far too much.

The "Short Chronicle" was only a list of names and dates but these dates were not justified. If Newton was to silence his critics he needed to establish his system with proofs. To this end he spent the rest of his life working day and night on the chronology. But he died with the chronology unfinished and the controversy was still raging in 1727 and continued to rage well into the nineteenth century.[112]

The Chronology of Ancient Kingdoms Amended

From the hundreds of unpublished manuscripts left by Newton after his death, apart from a couple of mathematical manuscripts, his literary executive Thomas Pellett found only two groups of works suitable for publication.[113] The first was a set of manuscripts on chronology and the second were two manuscripts on prophecies. Although Pellet claimed that the prophecies were imperfect, they were nevertheless worthy of publication. No publishers could be found to buy these works of prophecy, however, and they were finally prepared for press by Newton's nephew Benjamin Smith[114] and published in 1733 as *The Observations upon the Prophecies of Daniel and the Apocalypse of St John*. The set of manuscripts of Newton's justification for his chronology was compiled and arranged by John Conduitt and published in 1728 as *Chronology of Ancient Kingdoms Amended*.

As stated in the previous chapter, later biographies have assumed that his work on his *Chronology* was the work of an ageing Newton who had lost his taste and ability for science as a result of the nervous breakdown he suffered in 1693.[115] Although

1693 was indeed a black year for Newton, he did continue to research and add to the science of his day, albeit to a lesser degree than in the productive years of his youth. In 1695, he was appointed Warden of the Mint; in 1699 he became Master of the Mint overseeing the great recoinage of England and in 1703 he became the president of the Royal Society. The overseeing of the recoinage depended not only on Newton's excellent command of mathematics but also on vast organisational skills that entailed streamlining the mint by establishing country branch Mints[116] (White, 1998). In John Conduitt's "Memoirs of Sir Isaac Newton" he claimed that in his employment at the Mint Newton "had frequent opportunities of employing his skill in mathematics and chemistry, particularly in his table of assays of foreign coins, which is printed at the end of Dr Arbuthnott's book of coins".[117] Furthermore, the second edition of the *Principia* was revised and published in 1713 with a new improved version of his lunar theory.[118] These were not the works of an ageing and perhaps senile Newton and these works were in progress when he was attempting to complete work on the *Chronology*.

The *Chronology of Ancient Kingdoms Amended* is a curious and tedious book. David Castillejo claimed that the *Chronology*, on first reading, "is so thick and boring as to be almost impenetrable".[119] The book cannot be considered to be a success and is exceptionally dull. However, John Conduitt did compile and arrange it from unfinished manuscripts.

Newton attempted to put *Chronology* on a scientific footing and he approached chronology in a new way. Firstly, he provided an incisive critique of the ancient chronologists who he mostly rebuked for "their erroneous calculations and their ignorance of fact".[120] Secondly, he used proofs that were based upon astronomy to fix certain dates; these events were reported by the ancient chronologists, but Newton carefully and with extensive proof corrected their "poor and ignorant calculations".[121] Finally, he considered the entire literary evidence of the chronologies of the ancient kingdoms that he had collection from theologians, genealogists, poets, dramatists and from the Bible, which Newton considered to be the oldest and the most reliable document of all time.[122] Yet despite this apparent detail, one of the major criticisms of his work is his lack of proofs and references.

In the Preface to a 1770 edition of *Chronology,* which was in the form of a correspondence between Dr Hunt and the Bishop of Rochester, the Bishop reiterated to Dr Hunt the controversy that surrounded the publication of the "Short Chronology". He claimed that after Newton's death, sixteen drafts of the Chronology were found. The Bishop expressed his concerns about Newton's methods of writing:

> It is a pity, that he took so much of the same method in his chronology which he took in his Principia & c: concealing his proofs and leaving it to the sagacity of others to discover them. For want of these, in some instances what he says on chronology does not sufficiently appear at present to rest upon any thing but his assertions;. . . But proofs he may have had, which he chose to conceal, though what now stands in the Margin in those few places may have come from another hand, and may not amount to a full proof, as it pretends to do.[123]

The main theme of this strange and rather negative Preface for the book is that Newton did not finish the book himself and that he did not intend to publish it himself until the last years of his life and only then because he wanted to justify it in the light of the controversy surrounding this work. In the many years that he had worked on it and in the many redrafts of it, he made "few alterations in it, for the sake of shortening it. . ., and leaving out in every later copy some of the authorities and reference, upon which he had grounded his opinions".[124] The Bishop of Rochester claimed that it was he who had persuaded Newton to prepare the *Chronology* for publication to defend himself against the critics. However, the Bishop clearly blamed the editors and the number of drafts for the failures in the book.

Newton's insistence upon changing the dating of civilization is strongly proclaimed throughout the book. The "Short Chronicle" is in front of the book, followed by six chapters:

Chapter 2 – Of the Chronology of the First Ages of the Greek
Chapter 3 – Of the Empire of Egypt
Chapter 4 – Of the Assyrian Empire
Chapter 5 – Of the Two Contemporary Empires of the Babylonians and Medes
Chapter 6 – A Description of the Temple
Chapter 7 – Of the Empire of the Persians

Chapter 2 begins with "All nations, before they begun to keep exact accounts of time, have been prone to raise their Antiquities".[125] He claimed that the Greeks had "no public table or inscription older than the Laws of Draco".[126] Thus, according to Newton, the Greeks had no written records before the early seventh century BC. As for the Parian Marble acquired by the Earl of Arundel that had been hailed as the missing link of chronologies in the early seventeenth century, Newton dismissed it and stated that it had been composed 50 years after the death of Alexander the Great and the chronology had been reckoned backwards from that time. Also, the Greek chronologies claimed that the kings who had ruled before the times of the Persian Empire reigned for 35–40 years each. For Newton, in

> the ordinary course of nature kings reign, one with another, about eighteen or twenty years a-piece: and if in some instances they reign, one with another, five or six years longer, in others they reign is much shorter: eighteen or twenty years is a medium.[127]

In short, Newton claimed that the Greeks had doubled their antiquity.

Newton carefully assessed and compared the Greek chronologies and restructured the dates accordingly. He began dating the Greek chronology from two historical events; the first Olympiad, which he dated to 784 BC and more importantly the death of King Solomon which he dated to 982 BC. For example, he placed the return of Bacchus from India about 10 years after the death of Solomon[128]; the Argonautic expedition was 43 years after the death of Solomon[129]; the coming of the Danaus into Greece was about 40–45 years after the death of Solomon[130]; Prometheus had been left on Mount Caucasus 44 year after the death of Solomon[131]; the fall of Troy about 76 years after the death of Solomon[132]; the era of Nabonassar began 96 years after the death of Solomon[133]; the return of the Heraclides in Peloponnesus, 159 years after the death of Solomon,[134] the forty-ninth Olympiad was 397 years after the death of Solomon[135] and many others are dated from this point. Also the Greek hero Theseus was born in the thirty-third year of Solomon's reign.[136] Thus, the actions and events of the ancient Greek heroes were measured and dated by the death of a king of another nation.

It is not only the Greek Empire that Newton accused of exaggerating their antiquity. He claimed that the Egyptians, who had expanded their Empire eastwards to India and westwards to the Atlantic Ocean, "out of vanity have made this monarchy some thousands of years older than the world".[137] With the exception of the Old Testament, Newton was very circumspect on Egyptian chronology before Herodotus. He claimed that "Herodotus whose history the more I examine it the truer I find it, gives the best account of the ancient state of this nation".[138] But it was the Old Testament that was the oldest and the most reliable source for the Scriptures which "are by far the oldest records now extant"[139] and these Newton believed had been compiled from older books and records that were now lost.[140] There was "no instance of letters for writing down sounds before the days of David, in any other nation besides the posterity of Abraham".[141] He claimed that the written records of the Jews had been undisturbed until the fall of the Temple, but even with this disruption their records were relatively unscathed compared to the Gentile records.

The original religion, he believed, was revealed to the primordial people at Creation and this religion was preserved under the government of Noah and his sons. They were

> of one language, one society, and one religion: and then they divided the earth, being perhaps... forced to leave off building the tower of Babel: and from thence they spread themselves into the several countries which fell to their shares, carrying along with them the laws, customs and religion.[142]

This religion was passed through Abraham to the Jews. It was the religion of Moses and the prophets. "This was the morality and religion of the first ages, still called by the Jews, 'The precepts of the sons of Noah'." Furthermore "this is the primitive religion of both Jews and Christians, and ought to be the standing religion of all nations".[143] In his *Chronology*, this original religion was corrupted by kingship, and the veneration of kings was the source of superstition. In addition, the primitive religion of the Jews and Christians who worshiped the one God – this primitive monotheism – predated all Christian revelation; particularly, the revelations of the Trinity which Newton disagreed with.

For Newton, the Kingdom of Israel was the first significant political society that could be truly called a civilisation. However, Newton did have trouble justifying this against the accounts in the Old Testament, which he had claimed to be the oldest existing records of any civilisation. In the Book of Exodus, it clearly reveals Moses leading the Israelites out of a large and powerful Egyptian Empire. Yet Newton diminished it to a country of little city kingdoms that was not united into a great Empire until the reigns of Ammon and Sesac.[144] In the Short Chronicle, Newton dated Ammon 1034 BC[145] and Sesac 1014 BC[146] both postdate the reigns of David (1059 BC)[147] and Solomon (1019 BC).[148] According to these dates, Egypt was not an Empire until after the establishment of the Kingdom of Israel. Throughout the chapter on the Egyptian Empire, Newton again used the Death of Solomon as a point of reference in dating reigns of Egyptian Kings, but to a lesser extent than he had for dating the Greek Empire.

Despite the importance that Newton gave to the Kingdom of Israel and the key focal point for dating the Empires of Greece and Egypt being the death of Solomon, there is no chapter on the Kingdom of Israel, and the description of the Temple of Solomon is curiously placed between the chapters "Of the Two Contemporary Empires of the Babylonians and Medes," and "Of the Empire of the Persians". In the "Short Chronicle", the beginning of Solomon's Reign in 1019 BC and the founding of the Temple in 1015 BC are given as key dates.[149] Yet the beginning of the chapter is quite dismissive: "The Temple of Solomon being destroyed by the Babylonians, it may not be amiss here to give a description of that edifice".[150] The only link with the previous chapter, and rationale of its position, was the Temple's destruction, which Newton dated 588 BC.[151] The chapter consists of a brief description of its floor plan and is barely 3,000 words long, with three illustrated floor plans. There is no mention of the style of architecture, its splendour or its significance. The analysis lacks any enthusiasm and is a highly clinical description. This along with its brevity appears to contradict the importance given to Solomon throughout the Greek and Egyptian chapters. Furthermore, its architectural description has problems and there are parts that do not make structural sense.[152]

Stukeley noted his disappointment with the plan in the *Chronology*, claiming that Newton "has come pretty near my ground plan of the Temple of Solomon, but he gives no uprights".[153] Babson MS 0434 is the only surviving manuscript that clearly considers the architecture of the Temple in detail. There are a few other manuscripts that outline the Temple, but none of these give the detail that Babson MS 0434 does. Conduitt would have had possession of Babson MS 0434 as it came into the Portsmouth collection via his daughter. Yet it shows no notes or corrections by Conduitt, as do the other manuscripts, in the course of editing for the *Chronology*.[154] Furthermore, the floor plan of the *Chronology* is distinctly different from the floor plans of Babson MS 0434,[155] making it clear that it is not the text used or adapted for the *Chronology*.

At Christmas in 1725, at the height of the controversy over the "Short Chronicle," Stukeley recalled a discussion he had with Newton about Solomon's Temple. Newton showed him some drawings of the Temple,[156] and it is possible that this was Babson MS 0434 that Newton showed Stukeley; although in the course of this conversation, recalled by Stukeley, they discussed a style of architecture which is not mentioned in Babson MS 0434.

Stukeley claimed that although they had not discussed the details of the Temple, they both agreed that it was not like any other design. According to Stukeley, Newton claimed that Solomon's Temple was the oldest temple and it was the original model for all subsequent temples. He argued that the workmen on the Egyptian temples had come from Jerusalem and that they imitated the Temple of Solomon and also that the Greeks had borrowed their style of architecture from Solomon's Temple. Stukeley claimed that the style of the Temple was Doric and that Newton agreed with him saying that "the Greeks advanced it (Doric) into the Ionic and the Corinthian, as the Latins into the Composite".[157] But this is not stated in Babson MS 0434 or anywhere else in his manuscripts. According to Newton, the Temple of Solomon was the model for Herod's Temple, which was Corinthian and Doric. In Herod's Temple, this model was preserved in the colonnade of Solomon which was a part of the ancient structure.[158] Furthermore, Newton claimed that Herod had also maintained the magnitude of the columns, their number, and the triple row in the atrium of the Temple[159]; that he copied the details of what had been preserved of the Temple of Solomon. However, Newton did not state directly that Solomon used Corinthian columns, only that the model was preserved in Herod's Temple, but he clearly did not mention that was was Doric. Stukeley's reminiscences appear to support his own concept of architectural development rather than that of Newton.

Newton had placed Israel in the forefront of civilization and this civilization was the first to be built using all the science and art available. Although letters, astronomy and the trade of carpenters had been invented by the merchants of the Red Sea out of necessity, the Israelites had not only retained ancient knowledge, but also used these inventions to advance their civilization. For there had been no mention of the trade of carpenters or good architecture before Solomon sent for Hiram the King of Tyre to supply him with such artificers.[160] Newton confidently stated that "I meet no mention of sumptuous Temples before the days of Solomon: new kingdoms begun then to build Sepulchers to their founders in the form of sumptuous Temples; and such Temples Hiram built in Tyre, Sesac in all Egypt, and Benhadad in Damascus".[161] He had made the Temple the point of reference for all the temples to follow and subjugated the other "new" civilization to the Kingdom of Judea. There were no Temples before Solomon for "Temple began in the days of Solomon".[162] More importantly, in the understanding of Babson MS 0434, the Temple not only was the beginning of cultured civilization which retained ancient knowledge and wisdom, but was also a blueprint for Biblical prophesy.

Principia and *Prisca Sapientia*

In 1887, *The Times* wrote:

> It may be asserted without fear of contradiction that of all the anniversaries celebrated this year none is noteworthy more than the bicentenary of the publication of Newton's 'Principia'. No single work has ever been published which has exerted a more signal influence on science and on the progress of civilisation.[163]

Newton, however, believed that the advancement of civilisation had already been made by the ancients and had been subsequently lost. His role was to uncover the lost *Prisca Sapientia* or ancient wisdom. In the 1690s, Newton began to write some scholia (notes) which claimed that natural philosophy was "rediscovered" from ancient wisdom. He wrote:

> The most ancient opinion of the Philosophers was that the fixed stars stood motionless in the highest parts of the world, and that the planets revolved about the Sun beneath these stars; that the Earth likewise is moved in an annual course, as well as with a daily motion about its own axis, and that the Sun or hearth of the Universe rests quietly at the centre of all things. For this was the belief of Philolaus, of Aristarchus of Samos, of Plato in his riper years, of the sect of the

> Pythagoreans, and (more ancient than these) of Anaximander and of that most sage king of the Romans, Numa Pompilius. The latter erected a temple to Vesta, round in form, and ordained perpetual fire to be maintained at its centre, to symbolize the round shape of the Orb with the solar fire at its centre. It is very probable that the Egyptians disseminated this opinion, however, for they were the oldest observers of the stars. It seems that the Greeks, a race more given to philology than to philosophy, obtained this philosophy which was the oldest and soundest of all from the Egyptians and neighbouring peoples; and the rites of Vesta speak of the spirit of the Egyptians, who depicted their mysteries in rites, ceremonies and hieroglyphs far surpassing the understanding of the vulgar. Afterwards Anaxagoras, Democritus and several others taught that the Earth stands unmoved in the middle of the world.[164]

Although this Scholium was written for the second edition of the *Principia*, it was not published by Newton but appeared in the introduction of David Gregory's *Astronomiae Physicae & Geometricae Elementa* published in 1702.[165]

According to Newton, this "Copernican" theory had been known and taught by the ancient philosophers. Newton considered himself to be not only rediscovering the mathematical principles of the ancient philosophers in his *Principia,* but also rediscovering the ancient geometrical methods. In an intended preface for the second edition written in 1710, Newton attributed knowledge of universal gravitation to the ancient Chaldaean and claimed that Pythagoras imported it to the Greeks and Romans.[166] According to Newton, the role of Pythagoras in the transmission of ancient wisdom to the Greeks was an important one. Pythagoras had learnt from Egypt and Phoenicia the original and pristine religion of Noah, with all its knowledge of Natural Philosophy, including gravitation and the musical harmonies of the planets.[167] The intended preface reveals the relationship between Newton's "philosophical and mathematical classicism":

> The ancient geometers investigated things sought through analysis, demonstrated them when found out through synthesis, and published them when demonstrated so that they might be received into geometry. Once analysed they were not straightaway received into geometry: there was need of their solution through composition of their demonstrations. For the force of geometry and its every merit lay in the utter certainty of it, and that certainty in its splendidly composed demonstrations. In this science regard must be paid not only to the conciseness of writing but also to the certainty of things. And on that account I in the following treatise synthetically demonstrated the propositions found out through analysis.[168]

The geometry of the *Principia* adheres to the ancient methodological approach of presenting the theorems by synthesis.

In a draft for the second edition of the *Principia,* scholia were added to some of the propositions in Book III: The System of the World. In propositions IV–IX, Newton began with the gravitation of the Moon in its orbit to the inverse-square law generalised to include the all the bodies in the physical universe. Table 2.1 lays out these propositions along with the contents of the drafted scholium and the principal author cited by Newton.

In the draft of proposition VIII, Newton measured the distance of the planets through their ancient harmonic ratios:

> The ancients have not sufficiently explained by what proportion gravity decreases by moving away from the Planets. Yet they do appear to have adumbrated it by the harmony of the celestial spheres, by designating the Sun and the remaining six planets, Mercury, Venus,

Table 2.1 The contents of the drafted scholium[169]

Proposition	Content of the scholium	Principal author cited
IV – The gravitation of the Moon in its orbit	The Moon another Earth, like the other heavenly bodies	Plutarch, *De facie in orbe lunae, De placitis;* Diógenes Laertius; N Conti; Galileo
V – The mutual gravitation common to the planets and satellites	The various centres of gravity of the celestial bodies	Plutarch, *De facie in orbe lunae;* Democritus from Origen; Lucretius
VI – The gravity of bodies upon the planets is proportional to the masses of the same body	Proportionality of the mass and gravity	Lucretius; the atomists from Aristotle; Plutarch, *De placitis*
VII – Universal gravitation in the ratio of the masses of the heavenly bodies	Gravitational attraction and magnetic attraction	Plutarch, *De facie in orbe lunae;* Lucretius
VIII – The law of the inverse-square of the distances in the case of two bodies	The ratio between gravity and distance and musical scale	Pythagoras from Macrobius; Pliny; Proclus; Eusebius; Macrobius
IX – The law of inverse-squares generalized	Universal attraction and its cause according to the ancients	Thales from Diogenes Laertius; Pythagoras from Arisotle; the myth of Pan and the Orphic hymns from N. Conti; Macrobius

Earth, Mars, Jupiter, Saturn, by means of Apollo with the Lyre of seven strings, and measuring the intervals of the spheres by the intervals of the tones... For Pythagoras, as Macrobius attests, by stretching the intestines of sheep or the tendons of an ox by attaching various weights, and from this he learned the ratio of the celestial harmony. Thus, by means of such experiments he ascertained that the weights by which all tones which are of equal strings . . . were equal as the squares of the lengths of the string by which the musical instrument emits the same tones. [170]

The harmony of the spheres propagated by Pythagoras measured the distances between the planets. This harmonic measurement was in direct relationship with the inverse-square law. Thus, through Pythagoras's experimentation, the harmony of two strings when the tensions are equal is to the squares of their lengths. For Newton, this relation measured the distance and weights of the planet from the Sun, and thus the gravitational attraction between two planets could be understood in these terms. This ancient knowledge of natural philosophy had been lost over subsequent generations and Newton was attempting to recover this lost knowledge of natural philosophy.

Newton also practiced alchemy and studied ancient alchemical texts. Keynes MS 27 and 28 held at Kings College contains Newton's translations of and commentaries on the *Emerald Tablet* and *Seven Chapters,* two alchemical texts which are attributed to Hermes Benjamin Robert. In the seventeenth century, Hermes Trismegistus was considered to be an Egyptian priest who lived before Moses. His works contains the expressions such as "Son of God" and "the Father". Thus, Hermes Trismegistus was perceived as an important Gentile prophet who foresaw the coming of Christianity.[171]

The first of Newton's translations was the *Emerald Tablet* that dates to the early 1680s. It is one of the best-known alchemical texts. Betty Jo Teeter Dobbs has undertaken a study of Newton's translation and commentary. She claimed it was

> Of indeterminate but great antiquity, it was long supposed to encapsulate in its mysterious phrases all the occult wisdom of the ancients regarding divine actions in the creation of the world and regarding the alchemist's actions in the great work of alchemy, which was of course widely considered to be a little replication of divine creativity.[172]

In the early 1670s, prior to translating the *Emerald Tablet*, Newton experimented with sulphur and quicksilver. These experiments were conducted over a small furnace; he experimented with the heating, cooling and mixing of sulphur and quicksilver and recorded his experiments and observations. In one of his papers entitled "Of Natures Obvious Laws & Processes in Vegetation",[173] he claimed that the animals and minerals draw from the vegetable spirit the "material soul of all matter". All things are born, live and die that "ferment and principle of all vegetation". [174]

In the 1680s, Newton translated Hermes' text as:

> Tis true without lying, certain & most true.
>
> That wch is below is like that wch is above & that wch is above is like y^{t} wch is below to do ye miracles of one only thing.
>
> And as all things have been & arose from one by ye mediation of one: so all things have their birth from this one thing by adaptation.
>
> The Sun is its father, the moon its mother, the wind hath carried it in its belly, the earth is its nurse. The father of all perfection in ye whole world is here. Its force or power is entire if it be converted into earth.
>
> Separate thou ye earth from ye fire, ye subtle from the gross sweetly wth great industry. It ascends from ye earth to heaven & again it descends to ye earth & receives ye force of things superior & inferior.
>
> By this means you shall have ye glory of ye whole world & thereby all obscurity shall fly from you.
>
> Its force is above all force. For it vanquishes every subtle thing & penetrates every solid thing. So was ye world created.
>
> From this are & do come admirable adaptations whereof ye means (or process) is here in this.
>
> Hence I am called Hermes Trismegist, having the three parts of ye philosophy of ye whole world
>
> That wch I have said of ye operation of ye Sun is accomplished & ended.[175]

In Newton's commentary on this text he claimed sulphur and quicksilver were united; they acted on each other and are mutually transmuted into each other to create a nobler offspring.

> And just as all things were created from one Chaos by the design of one God, so in our art all things, that is the four elements, are born from this one thing, which our Chaos, by the design of the Artificer and the skilful adaptation of things. And this generation is similar to the human, truly from a father and mother, which are the Sun and the Moon. And when the Infant is conceived through the coition of these, he is borne continuously in the belly of the wind until the hour of birth, and after birth he is nourished at the breasts of foliated Earth until he grows up. This wind is the bath of the Sun and the Moon, and Mercurius, and the Dragon, and the Fire that succeeds in the third place as the governor of the work: and the earth is the nurse, Latona washed and cleansed, whom the Egyptians assuredly had for the nurse of Diana and

> Apollo, that is, the white and red tinctures. This is the source of all the perfection of the whole world. . . And just as the world was created from Chaos through the bringing forth of the light and through the separation of the aery firmament and of the waters from the earth, so our work brings forth the beginning out of black Chaos and its first matter through the separation of the elements and the illumination of matter. Whence arise the marvellous adaptations and arrangements in our work, the mode of which here was adumbrated in the creation of the world.[176]

Newton also showed interest in the thirteenth century Hermetic philosopher Ramon Lull. There is a small unfinished manuscript on the work by Lull, the first heading being "Ex Raymundi Seu de Quintessentiji" which is a set of notes on the separation of the elements. This is followed by "Ex Raymundi Lulli Libro Mercuriorum". Under this heading are some notes and the beginnings of an alchemical recipe, but it finishes in mid-sentence.[177] Further interest in Lull is shown in *Experimenta Raymundi*[178] and in the astronomico-alchemical symbolism of Lull and other Hermetic writers in *Tabula Smaragdina and Hieroglyphica Planetarum.*[179] Newton also translated[180] into Latin from French the Hermetic philosophy of Limojon de Didier's *Triomphe Hermetique* in a manuscript entitled *Epistola ad verso Hermetis Discipulos continens Claves Sex Principales Philosophiae Secretae.*[181]

Newton strongly believed in *prisca sapientia* and that the ancients had understood the true system of natural philosophy. However, he believed this knowledge had been lost through the subsequent corruption of the true religion. Since he believed that the *Emerald Tablet* was a text that was dated to the times before Moses, Newton considered it to be less corrupt and closer to the pure knowledge of the original religion of Noah. Because of this, such texts were significant both scientifically and theologically.

Science and theology came together because according to Newton's commentary on the *Emerald Tablet* in the alchemy of Hermes they have

> three parts of the philosophy of the whole world, since he signifies the Mercury of the philosophers, which is composed from the three strongest substances, and has body, soul, and spirit, and is mineral, vegetable, and animal, and has dominion in the mineral kingdom, vegetable kingdom, and the animal kingdom.[182]

Dobbs considered Newton's interest to be in the structure of matter and in proving that the science of alchemy demonstrated its changing form, and also that the theology of alchemy was its animating and activating spirit. There is also a striking similarity in the use of the word "dominion" to that of the General Scholium in the *Principia* with reference to God.[183] Newton closely associated the vegetable spirit that animated the mineral and the animal world with the light of the Sun – the hearth of the world.[184] His alchemical experiments supported his belief that the ancients such as Hermes had a perfect understanding of natural philosophy and that this knowledge had been lost with the corruption of the original religion.

In his General Scholium, Newton claimed that Numa Pompilius, the legendary second king of Rome and friend of Pythagoras, had erected a round temple to Vesta "and ordained perpetual fire to be maintained at its centre, to symbolize the round shape of the Orb with the solar fire at its centre".[185] For Newton this building preserved the ancient wisdom of the natural philosophy. In the architecture was encoded the wisdom of the ancients but this understanding had also been lost through time. However, this knowledge had also been recorded by the prophets of the Old Testament, Daniel and Ezekiel; and in the New Testament in Revelation of John the Divine. In the writings of these three prophets, the Temple of Jerusalem is a key element of their prophecies and the understanding and reconstructing of the Temple was part of the decoding of that knowledge.

Conclusion

Throughout the *Chronology*, the Kingdom of Israel clearly was the benchmark for understanding other civilizations. Its importance, however, is not recognised by its own chapter in the *Chronology*, but its importance is only acknowledge in relation to other civilisations, which makes the position and size of the chapter on Solomon's Temple even more curious, being placed after the civilisation that had destroyed it.

Many of the original manuscripts that became the *Chronology* are extremely difficult to read. Newton reused receipts and letters as paper for some of his manuscripts. The opening page of "The Original of Monarchies", one of the manuscripts that became *Chronology*, has writing going in four directions.[186] Many of the manuscripts are not ordered or dated and some were unfinished and fragmentary. The dating of the manuscripts is made more difficult by Newton's recycling of paper; he seemed to have kept paper for decades so the dates of the letters or watermarks can give no indication of the date of the manuscript.[187]

The placement of the chapter on Solomon's Temple after the Empires of the Babylonians and Medes in the *Chronology* does appear to be a curious decision by Newton. He sets out this chapter structure at the end of his life with a very reduced description of the Temple.[188] Babson MS 0434, which is a much fuller description of the Temple, has no additions and deletions by Conduitt that are evident in the other manuscripts used in his compiling of *Chronology*. However, although superficially there is a similarity of the floor plans of *Chronology* and Babson MS 0434, there are nevertheless significant differences. Elements are added in the plans of the *Chronology* that do not exist in Babson MS 0434. At the time of the publication of *Chronology* in 1728, the controversy over the Short Chronology was still raging and it is possible that Conduitt consciously decided to downplay the significance that Newton had given to the Temple of Solomon. However, the significance of the Temple of Solomon to Newton's philosophy is understated in the *Chronology* given its important in his manuscripts.

Newton's search to uncover ancient wisdom was not unique to him. However, Newton was not interested in just uncovering this lost knowledge – "his goal was the truth".[189] But his work on the chronology, ancient wisdom, alchemy and the Temple in his manuscripts should not be considered in isolation. His manuscripts on chronology and ancient wisdom were interlinked with his manuscripts on prophecy. In these manuscripts, the Temple takes an important role and the understanding of its structure and its architecture is of great significance.

Chapter 3
Prophesy and the Temple

Influences

Like chronology, ancient texts and alchemy, prophesy was a subject that Newton spent a great deal of time studying. In England, from the beginning of the Puritan unrest to the mid-eighteenth century, Apocalyptic and millenarian thought had became a widespread concern.[190] The work of English Millenarian Joseph Mede, *Clavis Apocalyptica* published in 1627, was to have a sustaining influence on notable scholars such as John Milton, Henry More, Isaac Newton and William Whiston. The majority of Apocalyptic and millenarian thought centred on the relationship between the Book of Daniel and the Book of Revelation. Mede's *Clavis Apocalyptica* was translated from Latin into English as *The Key of the Revelation* and was authorised for publication by the Parliament at the height of the Civil War in 1643.[191] Although Mede recognised that the correct interpretation of the prophetic language was paramount to understanding the prophet, he explored a synchronic scheme and did not lay down principles of Biblical hermeneutics.[192] Mede believed that the prophet spoke in a figurative and emblematic language that was once easier to understand than it was now. This language was after the tradition of the East, and Mede used non-Biblical sources to help in the understanding of the prophetic language.[193] Mede's pupil Henry More wanted to establish the prophecies of Daniel and John on a more scientific basis that was compatible with religious stability and a permanent social structure.[194] It did not matter whether the prophecies were fulfilled or what their meaning was, it was an understanding of the meaning of the language of the prophets, which must conform to rules and conventions, by which the meaning was conveyed.[195]

Newton's examination of the Biblical prophets dates back to his early years in Cambridge.[196] He owned books on prophecy and discussed the prophetic writings with More.[197] Newton owned all of More's published work on Daniel and the Apocalypse including *A Plain and Continued Exposition of the Several Prophesy or Divine Visions of the Prophet Daniel,* which was a gift to Newton from More. This book was heavily annotated by Newton with points of his disagreement.[198] Both More and Mede were significant influences on Newton. However, Newton's scheme of Biblical hermeneutics goes beyond More's and Mede's systems. Although Mede and More made mention of the Temple, in neither of their work is the Temple a central

T. Morrison, *Isaac Newton's Temple of Solomon and his Reconstruction of Sacred Architecture*, DOI 10.1007/978-3-0348-0046-4_3, © Springer Basel AG 2011

feature. Mede stressed the Tabernacle as being of greater importance.[199] He approached the Temple by considering the New Testament but he only included Chaps. 4 and 11 of Revelation and this choice of limited text excluded the architectural description of the Temple of Chap. 21. Mede claimed that the measuring of the courts in Revelation 11 "was propounded in the type of Angle measuring, in Ezekiel", and through this measuring

> You may see the place. But if any one shall conceive otherwise, and would rather refer the type of measuring to the signification of building, namely, that what the drawing or setting out of a model (that is the platform of a work to be done) is to builders the same as measuring to God in the language of the Prophets, it will come not withstanding to the same meaning.[200]

Through measuring and working out the plan of the building the builder was able to see how much work was to be done; similarly, in the prophetic language, God measured for the same reason. Mede supplied a woodcut of the floor plan or "plot" of the Temple which is exceptionally crude (see Fig. 3.1) and it is clear that he was not interested in the architecture of the Temple.

More is sometimes credited with influencing Newton's perception of the role of Jewish ritual in Revelation (Mede, 1643).[201] Despite this claim and having a close relationship with More (although not always in philosophical agreement),[202] with the exception of Joseph Mede, Newton did not look upon past interpreters favourably. He claimed that "all that I have seen beside the labours of Mr Mede have been so botched & framed without any due proportion, that I fear some of these authors did not so

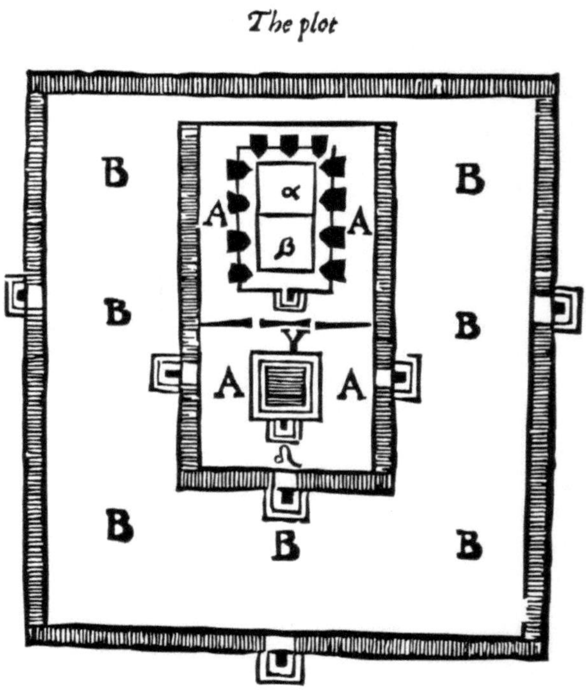

Fig. 3.1 Mede's floor plan or "plot" of the temple[205] (Drawn by the author from Mede, 1643, part 2, p. 6)

much as believe their own interpretation".[203] Newton acknowledged that "Mr Mede laid the foundation & I have built upon it".[204]

Newton believed that Mede had come closer to decoding the secrets of Revelation than anyone else. However, Newton conceded that the obscurity of the prophetic language was a deliberate device by the prophet to conceal the truth from the unfit or to delay the deciphering until the right time or the right person to decode them. At the end of his prophecy Daniel claimed, "and none of the wicked shall understand; but the wise shall understand".[206] About this Newton stated:

> To assign but one meaning to one place of scripture; unless it be perhaps by way of conjecture, or where the literal sense is designed to hide the more noble mystical sense as a shell the kernel from being tasted either by unworthy persons, or until such time as God shall think fit.[207]

The truth will be revealed by

> a remnant, a few scattered persons which God has chosen, such as without being led by interest, education, or humane authorities can set themselves sincerely & earnestly to search after [it].[208]

Only the wise and the worthy can interpret the prophet. It would only be those that were granted the tools of scientific method to reveal the workings of Providence in history who would be able to decode the prophets,[209] and Newton clearly thought that he was one of these worthy people.

Newton's only published work of prophecy was *The Observations upon the Prophecies of Daniel and the Apocalypse of St John*, posthumously published in two first English editions in London and Dublin in 1733 by his nephew Benjamin Smith. It was the result of several incomplete manuscripts. This work extended over 50 years but the published works was a small portion of Newton's writings on prophecy. *Observations* proved to be one of Newton's best sellers and it was also translated into Latin and German shortly after its first edition.[210] According to Richard S. Westfall, towards the end of his life Newton "had cleansed his *Observations*"[211] of any heretical material. It was further cleansed by the editor. Certainly the work that was published included carefully selected sections of Newton writings on prophecy, which had been highly edited and sanitised so that there is no reference to Newton's anti-Trinitarian beliefs. *Observations* was dedicated to Lord King, Baron of Ockham, Lord High Chancellor of Great Britain, who in his youth had shared Newton's anti-Trinitarian beliefs. King had also defended, without fee, Whiston, Newton's successor as Lucasian professor, at his trial for heresy for holding those same anti-Trinitarian beliefs, in 1713.[212] However, by 1725, when King had been raised to the post of Lord High Chancellor of Great Britain, he had forsaken his anti-Trinitarian beliefs. The heirs of Newton could publish without fear of Newton being exposed as a heretic. Nevertheless, there was an angry response from some of the more orthodox Anglicans who perceived Newton's work to have an anti-scriptural bias.[213]

Newton claimed that he studied the books of the Biblical prophets not to foretell the future but to reveal God and His creation. He believed:

> The folly of interpreters has been, to foretell times and things by this Prophecy, as if God designed to make them Prophets. By this rashness they have not only exposed themselves,

> but brought the prophecy also into contempt. The design of God was much otherwise. He gave this and the prophecies of the Old Testament, not to gratify men's curiosities by enabling them to foreknow things, but that after they were fulfilled they might be interpreted by the event, and his own providence, not to the interpreters. Be then manifest thereby to the world.[214]

It would be only after the prophesies had been fulfilled that their meaning would be made manifest.

The Language of the Prophets

The language of the prophets was figurative and often obscure but according to Newton they had a common language. He claimed, "John did not write in one language, Daniel in another, Isaiah in a third & the rest in others peculiar to themselves; but they all wrote in one & the same mystical language".[215] To understand the prophecy it was essential to understand the mystical and sacred language of the prophets and to this end it was necessary to understand the cryptic symbols that constituted that language.

Newton's interest in the development of a universal language dated back to 1661. It was considered that Latin was no longer a suitable language for the new study of natural philosophy, and a language that consisted of mathematics or ideograms may be more suitable for its expression and could be developed into a full language capable of expressing what Latin could not.[216] In a manuscript titled *Of an Universall Language*, he developed linguistic topics and his word list consists of 2,400 entries. This study is dated 1661; it was never completed[217] and there appears to be no later manuscript on this topic. It was not until about 1680,[218] when he began to turn to deciphering the language of the prophet, that he used a similar system of fixing a signification to types and phrases. He claimed:

> The rule I have followed has been to compare the several mystical places of the scripture where the same prophetic phrase or type is used & to fix such signification to that phrase as agrees best with all the places . . .[219]

This mystical and figurative language is taken from the natural world, the world of kingdoms and empires, and the political world.[220] The events of man were signified by signs from nature, an "analogy between the world natural and the world politique for the mystical language is founded in this analogy & will be best understood by considering the original".[221] In this he followed Mede, who claimed, "In the prophets . . . every kingdom and body of government resembles the world: as the parties also, the heaven, the earth, the stars serve for that representation".[222] However, Newton developed this concept into a mystical language. This mystical language, divined from nature, was the hieroglyphic language of the Egyptian priests and understood by the interpreters of dreams and the Chaldaean paraphrase.[223] Newton wrote:

> Now in heaven the Sun & Moon are by Interpreters of dreams put for the persons of Kings & Queens, but in sacred Prophesy which regards not single persons, the Sun is put for the whole species & race of Kings in the Kingdom or Kingdoms of the world politick shining with regal

> power & glory: the Moon for the body of the common people considered as the Kings wife: the stars for subordinate Kings Princes & great men or for Bishops & Rulers of the people of God when the Sun is Christ: Light for the glory truth & knowledge wherewith great & good men shine & illuminate others: darkness for obscurity of condition & for error & ignorance: darkening, smiting or setting of the Sun Moon & stars for the ceasing of a kingdom or for the desolation thereof proportional to the darkness: darkening the Sun turning the Moon into blood & falling of the stars for the same. New Moons for the return of a dispersed people into a body politique or ecclesiastique.[224]

The prophets, Newton found, used a mystical language, a language of dreams, not because the prophets were dreaming but because through this language the they were able to reveal God.[225] There was a correspondence between the symbols of heaven and earth. Using this signification, Newton derived 70 hieroglyphic images[226] (Mamiani, 2002), which were constructed according to the laws of grammar to form a language.[227]

To decode this language of hieroglyphic images, Newton formulated a set of rules to assist in the inscription of the prophetic language. He devised 16 rules for the interpretation of the Scripture.[228] Matania Z. Kochavi has simplified and summarised these 16 rules into 4 major points: (1) The entire prophetic text must be treated as one homogenous structure; (2) The entire text must be decoded in minute detail; (3) The interpretation of prophetic revelation must be simple and (4) The interpreter of the Prophetic text must interpret the text with the aid of historic events.[229] Newton's work on chronology testified to the fulfilment of the prophecies. In his writings on prophecy he scrupulously applied these rules.

For Newton, the truth of the message of God was in the mystical and cryptic hieroglyphs in the prophecies. "All sacred Prophesies are given for the use of the Church, & therefore they are all to be understood by the Church in those ages for whose use God intended them".[230] As the prophesies were revealed to be true this confirmed the truth of God's word. If the message of God were to be revealed, the oldest prophet Daniel was the "easiest to be understood: and therefore in those things which relate to the last times, he must be the key to the rest" (Newton, 1999).[231] For Newton, this "proof" of God's message was highly significant for to reject Daniel's prophecy, "is to reject the Christian religion, for this is founded upon his prophecy concerning the Messiah".[232]

But a reliable analogy of the depiction in the prophecies and its earthly realisation depended on the Biblical text not being corrupt. In *Observations* Newton dedicated a chapter to the authenticity of the Old Testament. The prophet's writings contain the covenant between God and his people. Despite the vagaries of how the Old Testament was transmitted through time he believed that the Book of Daniel had not been corrupted. Newton claimed, "The Book of Daniel is a collection of papers written at several times. The six last chapters contain Prophecies written at several times by Daniel himself while the six first are a collection of historical papers written by others".[233]

In the second part of *Observations* on the Revelation of John the Divine, a chapter is dedicated to the authenticity of the New Testament and here Newton was a lot more critical. Saint Irenaeus (d.c200) had dated Revelation in the time of the Emperor

Domitian (51–96) and other theologians agreed with this dating which confirmed that John's Revelation was disconnected from Christ's life on earth. But Newton claimed that Revelation was the earliest book of the New Testament and it was written before the destruction of the Temple of Herod in the year 70. Newton claimed that this was confirmed "by the allusions in the Apocalypse to the Temple and Altar, and holy City, as then standing".[234] As the oldest book in the New Testament, it was the earliest Christian text.

Newton believed that the Book of Daniel and Revelation demonstrated how the corruption of the Christian Church had been foretold in the Biblical prophecies and this had been proven by the historical development of the Church.[235] The majority of *Observations* is dedicated to a very complex chronology of the historical development of the successive kingdoms and rules which are contemporaneous with the events and depiction of the prophecies. This prophetic chronology spans from the fall of the Temple of Herod in Jerusalem, destroyed by the Romans in 70 AD, to Newton's own time.

His belief that prophesy was revealed though hieroglyphs, which contained historic truth as a collaboration of the natural world and the political world, corresponded with the earthly rituals and parts of the Temple of Jerusalem and the Celestial City of the future.[236] The Temple, ritual objects and rituals that were preformed within the Temple appear in both the prophecies of Daniel and John. In Daniel, God commanded him to shut and seal the book which contained the names of the people. This book would remain sealed until the end of time when they "shall awake, some to everlasting life, and some to shame and everlasting contempt".[237] Newton believed that John's prophecy

> is called the Revelation, with respect to the scripture of truth, which Daniel was commanded to shut up and seal, till the time of the end. Daniel sealed it until the time of the end; and until that time comes, the Lamb is opening the seals.[238]

The seals that Daniel closed were opened by the Lamb in Revelation at the end of time. In the Revelation of John this book was sealed with seven seals and each is unsealed one by one marking the end of time.

In the opening of Babson MS 0434, Newton stated:

> Generally it is agreed that the future is foretold in the legal constitutions (the Hebrew texts) and the Apostle Paul, who bears witness to this in his writings, Colossians 2.17 and Hebrew 9.23. Thereupon these constitutions are more suitable than the natural world from which the prophets might choose the figures, and the *Apocalypse* is full of this sort of figures and these constitutions and those of the *Apocalypse* are thus like twins, since they prophesy from the same two matters, they explain themselves mutually, they may not be understand apart. It is in fact a legal and sealed up book (The Torah) at hand for Him who is seated of the throne and its seals are undone in the Apocalypse. Consider the world-universe of the Israelites and the meaning of its parts and the significant of its ceremonies, which need to be explained.[239]

The Books of Law and the Apocalypse, the opening of the seals, are explained mutually and for Newton "the Temple is the scene of the visions"[240] of the prophets.

However, this much edited version of Newton's work on prophecy, *Observations,* has many passages omitted. The manuscript *The First Book Concerning the Prophets* comprises two treatises on prophecy, the first written circa mid-1680s and the second, a redraft with many additions, written around 1705–10. Both have the synopsis of prophetic images or figures which is very similar to Chapter Two, Part One of *Observations*, but in *Observations* there is a notable omission. In the first draft of *The First Book Concerning the Prophets*, Newton claimed:

> In the Apocalypse the world natural is represented by the Temple of Jerusalem & the parts of this world by the analogous parts of the Temple: as heaven by the house of the Temple; the highest heaven by the most holy; the Throne of God in heaven by the Ark; the Sun by the bright flame of the fire of the Altar, or by the face of the Son of Man shining through this flame like the Sun in his strength; the Moon by the burning coals upon the Altar convex above & flat below like an half Moon; the stars by the Lamps; thunder by the song of the Temple & lightning by the flashing of the fire of the Altar; the earth by the Area of the courts & the sea by the great brazen Laver. And hence the parts of the Temple have the same signification with the analogous parts of the world.[241]

This passage was written around the same time that Babson MS 0434 was written.

Thirty years later it is redrafted thus:

> Sometimes a body politick is represented by the Building of a City or Temple . . . if it be a Temple the parts hereof have the same signification with the analogous parts of the World, for Temples were anciently contrived to represent the frame of the Universe as the true Temple of the great God. Heaven is represented by the Holy place or main body of the edifice, the highest heaven by the most Holy or Adytum, the throne of God by the Ark, the Sun by the bright flame of the fire of the Altar or by the face of the Son of man shining through this flame like the Sun in his strength, the moon by the burning coals upon the Altar convex above & flat below like an half moon, the stars by the lamps, thunder by the song of the Temple, lightning by the flashing of the fire of the Altar, the Angels or inhabitants of heaven by Cherubim carved round the temple, the Sea by the great brazen laver, the earth by the area of the Courts & the bottomless pit or lower parts of the earth called Hades & Hell by the sink which ran down into the earth from the great Altar & was covered with a stone to open & shut. And all these parts of the Temple have the same signification with the parts of the world which they represent. And in allusion to the River Siloam which ran by the Temple of Jerusalem & flowed thence eastward & was by the Jewish Doctors accounted a type of the spirit, a River of life flowing eastward from the throne of God with trees of life growing on the banks thereof is put for the Law of God going out from the Throne of the kingdom to the Nations, the fruit of the trees & the water of the River being that spiritual meat & drink which Christ has represented by his body & blood & by the bread & wine in the Eucharist; & which were also prefigured by the Manna & rock of water in the wilderness.[242]

The Practices of the Prytanæa and the Temple

In the manuscript *The Original of Religions*, which unfortunately is undated, Newton examined the ancient religious practices of the Prytanæa, which he believed was the original religion that had been derived from Noah and his sons.[243] The Prytanæa understood the mathematical principle of God's orderly design that sustained the solar system. Newton perceived that they had a pure knowledge of the workings of the structure of the universe. They practiced the ritual around a sacred fire, preserving the divine wisdom of the heliocentric universe. For Newton,

> as the Tabernacle was contrived by Moses to be a symbol of the heavens (as St. Paul & Josephus teach) so were the Prytanæa amongst the nations. And as the Tabernacle was a

> symbol of the heavens, so were the Prytanæa amongst the nations. The whole heavens they reckoned to be the true & real Temple of God & therefore that a Prytanæum[244] might deserve the name of his Temple they framed it so as in the fittest manner to represent the whole system of the heavens. A point of religion then which nothing can be more rational. . .the fire in the middle of the Prytanæum was taken for a symbol of the centre of the world.[245]

All parts of the Temple represented the political world, which was the microcosm of the universe, designed by God. Moses, in erecting his altar and the Tabernacle, preserved the religion of his ancestors by placing the fire in the common centre of the court of the Priests "so as to make it a symbol of the world".[246] Moses possessed the secret knowledge about the structure of the universe. "Solomon maintained the proportions of the areas of Moses in the construction of the Temple, but he doubled the measurements".[247] Newton established that the Temple of Solomon was the model of all temples. Thus, this was the model microcosm of the universe and revealed the mind of the Supreme Architect – the mind of God. Newton told Conduitt that he believed that an analysis of the origins of religion would solve the problem of religious disputes just as his *Principia* had solved the problems of natural philosophy.[248] He claimed:

> So then was one design of the first institution of the true religion to propose to mankind by the frame of the ancient Temples, the study of the frame of the world as the true Temple of the great God they worshipped. And thence it was that the Priests anciently were above other men well skilled in the knowledge of the true frame of Nature & accounted it a great part of their Theology.[249]

The Temple was also the stage or scene of the Apocalypse on earth which corresponded with the Temple of God which was open in heaven. Newton explained the opening of the seven seals in the Apocalypse of John through the ceremonies of the Temple. The Temple was the scene of the prophetic visions, and the visions in the Temple corresponded to the Jewish feast of the seventh month. In Revelation 1:20 there are seven stars which are the seven angels of the seven churches. Newton claimed,

> And the seven angels were also Chief Priests because they came out of the Temple where none but Chief Priests enter, & were clothed in pure & white linen & had their breasts girded with golden girdles which is the Priests habit, & at seven sacrifices poured out seven Vials or drink offerings & sounded seven trumpets. And as they are Priests so they are said to be before the throne or Adytum of the Temple & are considered in the Apocalypse as having the oversight of all things, being called the seven eyes of the Lamb, & the seven spirits of God sent forth into all the earth, that is, the seven messengers of God. Angels signify messengers & are put in general for officers & ministers of the Temple & by consequence the seven chief Angels for the seven chief Officers.[250]

In Revelation 4:4, around the Throne of God sit 24 Elders. For Newton, "These Elders are the Priests and Levites divided into twenty-four courses under twenty-four Princes who had twenty-four chambers about the Temple, twelve on one side of the Priests court and twelve on the other side thereof".[251] He examined the vision of John and each of the events in the Apocalypse and positioned them in the Temple. For example:

> "And out of [the] throne proceeded lightning and thunderings, and voices" (Revelation 4:5) viz. the flashes of the fire upon the Altar at the morning-sacrifice, and the thundering voices of these that sounded the trumpets, and sung at the Easter gate of the Priest's Court . . . "And

> before the throne was a sea of glass clear as crystal" (Revelation 4:6); the brazen sea between the porch of the Temple and the Altar filled with clear water.[252]

Newton represented the people of Israel with the hieroglyphs of the Cherubim and the Seraphim. The Cherubim show the presence of God's Glory in the earthly sanctuary; their celestial counterparts are the Seraphim who are God's servants in a Heavenly sanctuary. The Cherubim have one body and four faces; a face of a lion, a face of an ox, a face of a man and a face of an eagle.[253] For Newton, the lion, the ox, the man and the eagle were the four standards of the legion of the Jewish tribes: the three tribes under the standard of Judah were represented by a lion; the three tribes under the standard of Ephraim were represented by an ox; the three tribes under the standard of Reuben were presented by a man and the three tribes under the standard of Dan were presented as an eagle.[254] Newton took the symbol of Dan to be the eagle when it is in fact a scorpion (Newton, 1999).[255] By equating the faces of the Cherubim with the tribes of Israel and replacing the scorpion with the eagle he was representing the tribes of Israel by the symbols of the four evangelists. These symbols are also replicated in the Seraphim. The four beasts of the apocalypse were the "four Seraphim standing in the four sides of the peoples court:"[256] the one in the east has the head of a lion; the one in the west has the head of the ox; the one in the south has the head a man and the one in the north has the head of the eagle.

Seven days before the feast of the seventh month the Chief Priest would continuously study the Book of Law in the Temple.

> There were certain priests appointed by the Sanhedrin to be with him those seven days in one of his chambers in the Temple, and there to discourse with him about the Law, and read it to him, and put him in mind of reading and studying it himself. This his opening and reading the Law those seven days, is alluded unto the Lamb's opening the seals.[257]

The priests used the parts of the Temple that Newton referred to in his reconstruction in Babson MS 0434. Newton carefully justified these measurements and the position of these chambers within the Temple. The position and measurement of the parts of the Temple were important to the whole concept of the Temple.

In Revelation 11:1–2, John was commanded by the Angel to "Rise and measure the Temple of God and the Altar, and them that worship therein. But the court which is without the Temple leave out, and measure it not; for it is given unto the Gentiles". Newton claimed that the Temple plus the court of the Gentiles referred to Ezekiel's measuring the Temple of Solomon but that the measuring of the "Temple of God and the Altar, and them that worship therein, without the Gentiles court signified the building of the second Temple".[258] Solomon's Temple was the first Temple, representing the history of the Jewish people, while John's prophesies were of the second Temple which was symbolic of the history of the Christian community.

According to Newton, on the seventh day of the feast the Angel/Chief Priest sounded the seventh trumpet that alluded to the destruction of the first Temple and the Babylonian captivity, followed by the building of the second Temple by Zerubbabel. This Temple was not shown to John for the outer court was not rebuilt but given to the Gentiles, and this marked the beginning of the corruption of the original church.

> For measuring is a type of building (Zech. 2 & Ezek. 40) & the outward court of the Temple was not rebuilt by Zerubbabel but left open to the Babylonians & called the court of the Gentiles. Measuring is also a type of distinguishing that which is measured from that which is left unmeasured or from that which is measured for another purpose...[259]

While those of the 12 tribes who remained in the measured Temple "built up a new spiritual house, an holy priesthood, to offer up spiritual sacrifices, acceptable to God by Jesus Christ",[260] those who stood in the unmeasured outer court received the mark of the Beast, and those "worshipping the Beast & his Image as the heathens worshipped their false Gods & Idols, are called Gentiles".[261]

In Revelation 11:19, John saw the Temple of God open and there was seen in this Temple the Ark of the Covenant. Newton argued that since the Temple contained the Ark that it was the first Temple. However, when the Angels/Chief Priests poured out the seven vials of wrath that destroyed the Earth, that destruction was signalled by the seventh trumpet, which came out of the second Temple for it had no courtyard. The prophesy of the Book was represented by the Book of the Law. The Chief Priest read aloud from the Book of Law in the Temple and this was therefore repeated and interpreted in John's vision as the prophesy of the Apocalypses which began with the Temple of God opening to Heaven and ended with the sound of the seventh trumpet.[262]

The original religion was that of the ancient religion of Prytanæa, those who followed the religion of Noah, who understood the mathematical principle of God's orderly design that sustained the solar system. Their practiced ritual around a sacred fire, which preserved the divine wisdom of the heliocentric universe, was followed by Moses and in turn by Solomon. The laws of nature were the Laws of God. Yet these laws were not static. Descartes had presented a view of the universe that was a completely mechanical universe, which once set in motion would move in perpetual movement around the sun. But by the 1660s there were serious misgivings that this system excluded any role for God.[263] Although Newtonian mechanics did reinforce this notion of a celestial clockwork universe Newton noted small variations in the orbits of Jupiter and Saturn.[264] Also, in *Opticks,* Newton claimed that Comets could not keep their orbs.[265] The universe was in decay.[266] Newton concluded that God had to intervene occasionally to "repair" and "restore" the balance of the universe. "Newton and many of his English contemporaries seem, like the Stoics, to view the cosmos as going through successive cycles. The destroyed Earth of the next cycle would emerge".[267]

This cyclical nature of decay and restoration was emphasised by Newton's choice of the feast of the seventh month to be celebrated in the Temple. The Biblical name for this feast is "Yom Teruah, which means 'the day of the awakening blast'".[268] It is commonly called the "feast of Trumpets" and not only celebrates the resurrection of the dead but also the Jewish New Year. In the *Chronology,* Newton claimed that the Israelites used a lunar–solar calendar and when the year fell behind it was just before the feast of the seventh month that an inter-calendar month was added.[269]

Newton believed that this cycle of decay of the universe was linked with the history and corruption of the Church and was foretold by the prophets Daniel and John.[270] The macrocosm was encoded in the Temple and this building and the hieroglyphs were a key to understanding the prophets. In Yahuda MS 1.1, Newton referred to his hieroglyphs as constructions (plural) according to the laws of grammar.[271] However, he also referred to the "construction of the Apocalypse"[272] implying a singular structure. Within this context, the measuring and construction of the Temple to strict and justified rules forms a link between the construction of the Apocalypse and the Temple. This is also true of the construction of the Apocalypse which consisted of constructed hieroglyphs. Similarly the Temple contained hieroglyphs in the form of the rituals performed by the Chief Priests.

In Babson MS 0434. the Temple's plan, architecture and the function of the chambers and courts are carefully defined. In this structure are encoded the laws of nature and the places where the rituals are performed. But above all, everything was measured and Newton checked those measurements against a wide range of ancient sources.[273] "For measuring is a type of building" it defined those who kept faith and followed the line of the original religion and the unmeasured defined its corruption. The measuring and the ritual performed by the Chief Priest herald in the stages in the evolution of the Church and historical events that lead to that evolution. The cycles of decay and renewal of this history are aligned to the cycle of the universe encoded into the Temple.

Is Ezekiel's Vision of the Temple the Same as Solomon's Temple?

Richard Westfall[274] and Ayval Leshem[275] considered Solomon's Temple and Ezekiel's Temple to be two different Temples. This was a highly debated question in the seventeenth century.[276] The debate had originated in *In Ezechielem Explanationes et Apparatus Vrbis Templi Hierosolymitani (Ezekiel's explanation and the preparation of the cities and of the temple of Jerusalem)*, written by Juan Battista Villalpando and published in 1604. Villapando claimed that Ezekiel's vision was a vision of Solomon's Temple.[277] Newton, who had studied Villalpando's reconstruction,[278] concurred that they were the same Temple and Newton's main source for his reconstruction of Solomon's Temple was the Book of Ezekiel. William Whiston claimed:

> As for Sir *I.N'*.s description of Solomon's Temple; [I think he should call it Ezekiel's Temple; for he takes it principally from Ezekiel, who describes neither Solomon's, nor Zorebabels', nor Herod's, but the Jews future Temple] I reserve its examination till I publish my own plan of all those Temples.[279]

Through Newton's use of Ezekiel as his main source for the structure of Solomon's Temple he emphasised the measurement, which he checked against various other sources, which highlighted the measurement in Revelation. The Prophet Ezekiel's vision became the "construction of the Apocalypse" or at least the end of the decay and the beginning of the renewal.

Whiston designed and constructed a model of Ezekiel's Temple in 1726; unfortunately no details of the models have survived, and "he used these models as the basis for his millennial lectures about the recall of the Jews, the rebuilding of the Temple, and the second coming".[280] After Newton's death, Whiston used the Newtonian methodology of Scriptural interpretation to reveal that current early events were "leading up to the millennium and the physical reconstruction of the Temple in Jerusalem".[281] Whiston believed that the draft copies of Newton's work on prophecy were much more coherent than *Observations,* and he claimed that most of it was written when Newton was in his forties and at most in his fifties.[282] Although Newton revised his papers on prophecy in his old age, in general the revision sanitised what had already been written. The main ideas and structure of *Observations* were conceived and written at the same time as he was working on the *Principia.*

However, although Whiston utilised the Newtonian methodology of Scriptural interpretation, he was more critical of Newton's work of chronology and prophesy than supportive. He attacked the foundations of Newton's research rather than his methodology, which he supported. He claimed that Newton was unhappy with all the histories, chronologies and prophesies that he had collected from ancient times. Consequently, Newton

> ventured to furnish himself with new materials from his own sagacity: take not a few things for granted as the foundation of his reasonings, which best suited his preconceived opinions tho' they were not supported, nay tho' they were even contradicted by a great deal of real and positive evidence. This way of supposals without nay, against positive proofs runs through Sir Isaac Newton's whole chronology: as I have at large shewed in my consultation of it. And this is but too frequent in his observations on the prophecies.[283]

Whiston maintained that Newton was led into making "many mistakes in both these treaties". He challenged Newton's sources and asserted that although the Temple services were frequently alluded to in the Apocalypse, and that both Mede and Newton agreed that they are an important part and are of very great value; "yet does it seem to me that Sir Isaac Newton has therein carried this matter farther than it will bear: and farther than any good evidence can support him".[284] Despite his criticism, Whiston believed that Newton's "discoveries" in prophesy were significant and should be considered and discussed "because the momentous *truths* here laid down by so great a man have a right to be supported and recommended; and the public has a right to have his *mistakes* noted and corrected".[285]

The *Observations* stimulated other commentaries and defences of Newton's work. However, most of these works ignored the significance that Newton placed on the Temple. John Saint Clair's *Observations on Certain Passages in Daniel and the Apocalypse of John* published in 1755 is a defence of both Mede and Newton's work on Prophecy. Yet the Temple is given no importance; he dealt with Revelation 21 by stating that it was "more convenient beholding the City, rather than to shew that Christ's Church was established on a mountain. You find the like in Ezekiel".[286] Many contemporary writers ignored the Temple entirely.[287] For although the published works of the *Chronology* and the *Observations* do give the Temple an important role, the real significance of that role in Newton's thought is seen only in his unpublished papers. These papers make it clear that the role given to the Temple by the editors of the *Chronology* and the *Observations* was disproportionately minor as compared to the role given to it in Newton's unpublished papers.

Conclusion

The title of Babson MS 0434, *Introduction to the Lexicon of the Prophets, Part two: About the appearance of the Jewish Temple,* implies that it should have been included in the *Observations.* Yet apart from Newton's the opening paragraph quoted above, it is unlike the rest of his prophetic work. Ezekiel's vision of the Temple is discussed purely through its structure and its use. There is no mention of its prophetic meaning. Although Newton claimed that Ezekiel's description was the best in terms of his analysis of the Temple, he also noted that Ezekiel could be obscure,[288] and that there were areas in the Temple that Ezekiel had not seen.[289] Furthermore he suggested that Ezekiel gave interpretations that were erroneous[290]; his narrations were extremely difficult to understand[291]; he left out descriptions that are described by others[292] and he omitted relevant details.[293] Despite his reservations regarding Ezekiel, in Babson MS 0434 Newton emphasises the measurements, the plan and how the building was used. Yet the title indicates that it was meant to be part of his lexicon of the Prophets. In a manuscript, *Treatise on Revelation* dated mid 1680s (Newton, c mid-1680s),[294] a table of contents is given for a proposed structure for *The First Book Concerning the Language of the Prophets* consisting of five books but omitting Book Two. There are 10 titles of chapters in the first book and the tenth chapter is titled "Of the parts of the Temple".[295] However, there is no Book Two; after "Of the parts of the Temple", is the list of chapters for the third book. This leaves the question "was Book Two going to be the reconstruction of the Temple"? The *Treatise on Revelation* manuscript is dated around the same date as Babson MS 0434, so it does make this a possibility.

The omission of the plan of the Temple in *Observations* is equally as puzzling. In his unpublished manuscripts on prophecy, it is made clear how important the plan of the Temple is to the prophesy of the history of the Christian community. Yet the plan of the Temple has remained separate in Babson MS 0434. It could not have been easy for Smith to compile this work from the manuscripts left by Newton. However, Babson MS 0434 is not an isolated manuscript; it is part of Newton's overall system that was outlined and heavily edited by Conduitt and Smith in the *Chronology, Observations* and in his unpublished manuscripts. The Temple is "the scene of the visions" of Daniel and John and as such its meaning and construction need to be considered carefully and both the meaning and construction is seen in his unpublished papers.

Chapter 4
Prytanæum and the Floor Plan of the Temple of Solomon

The Prytanæum as the Frame of the Universe

In the *Chronology*, the corruption of the original religion is not expanded into any substantive conclusion. However, in his unpublished manuscripts, Newton expanded the subject of the ancient corruption of the original religion which led to the corruption of the Church in Newton's own time. Newton redrafted his work *Theologiæ Gentilis Origines Philosophicæ*[296] many times and left various tables of contents.[297] Although this indicates that he intended it to be published, the contents of these manuscripts were heretical and it seems unlikely that he would have intended it for publication in his lifetime and neither did his heirs consider all of the contents publishable. The original *Origines* consisted of rough drafts and notes, but he continued to develop and redraft the concepts of *Origines* from the 1680s until his death in 1727 in manuscripts such as *The Original of Religion*[298] from the early 1690s and *Irenicum*,[299] which is undated. These concepts were still being worked on when he was working on his *Chronology*.

Newton believed that the ancient Egyptians practiced a sacred philosophy that originated from knowledge of the stars – "Sacred philosophy obviously flourished in Egypt and was founded on the science of the stars".[300] He made the celestial iconography explicit in a description of a religious procession.

> In this procession the hymns of the first priest are associated with the harmony of the heavenly spheres. Next comes the astronomer with the holy books concerning the study of the stars. Next comes the sacred scribe who understands the planets, stars and the sacred things. Finally the priest and chief appear, who know all things as a consequence of studying the sacred rites and theology, and who close the entire procession. By bringing together a knowledge of the stars and the earth with the study of that which was the most important, Egyptians indicated that their theology concerned the study of the stars. Indeed, the gods of the Egyptians were stars and elements (Knoespel, 1999).[301]

The study of the stars was essential to the rituals of the priests. In *Origines,* Newton does not discuss the Prytanæum, but in the mid-1680s he was also working on Babson MS 0434, and there is a distinct change in emphasis in his redrafts. In the latter developments of *Origines*, the iconography of the building becomes essential to the enacting of the rituals. For Newton, "The religion most ancient and most generally received by the nations in the first ages was that of the Prytanæa or Vestal

T. Morrison, *Isaac Newton's Temple of Solomon and his Reconstruction of Sacred Architecture*, DOI 10.1007/978-3-0348-0046-4_4, © Springer Basel AG 2011

Temples".[302] This form of worship was spread by Noah and his sons, from Egypt, and at the heart of the Noachian religion lies the symbol of the Prytanæa – the sacrificial altar, the hearth, the centre of worship – which represented the structure of the heliocentric solar system. Although Noah kept the true religion, he and his sons were worshipped as gods and eventually as planets. Noah was worshipped as Saturn and his son Ham as Jupiter, and Ham's sons became identified with Hercules, Osiris, Antaeus and Busiris. In Egypt and elsewhere, the original religion became corrupt, with idolatrous worship of kings who were transmogrified figures of Noah and his family. Noah performed rituals of sacrifice around a sacred fire. Newton claimed:

> Noah & his sons carried with them the sacred fire from the Tower of Babylon into the land of Shinar . . . & Abraham carried it with him to offer Isaac & Æneas carried it with him from Troy & the ancient Kings of Greece & Persia carried it along with them into the field when they went to make war: so the sons of Noah when they went from him into their several countries took this fire along with their several families & the like was done by their sons & grandsons as oft as they went with their families to live at any considerable distance from one another in a distinct polity. And by this means I conceive it came to pass that the sacred fire at the first plantation of the earth was to be found in every City, as an essential part of the government, for in the first ages when the whole world was distinguished into as many kingdoms as cities.[303]

The original religion had continued with Abraham, but there were continuous external influences that tainted the religion. The Israelites understood the Prytanæa of the neighbouring nations, "which the Israelites should introduce into their Land, & therefore these Prytanæa were used in the Cities of Canaan & Syria before the days of Moses".[304] God sent Moses to teach the uncorrupted Noachian worship to the Jews. Moses taught the Jews no other than the true religion that was purged of the corruptions of the nations, who had added the idolatrous elements. Both Noah and Moses

> kept a perpetual sacred fire in a consecrated place for sacrifices. And as there was but one Prytanæum or Temple in the kingdom of the Jews so in the first kingdoms of the Nations so there was but one fire in a kingdom. When every city was a kingdom there was a Prytanæum in every City. When many cities united under one common council & thereby grew into one kingdom, there was in the chief city where the Council met a Prytanæum of a nobler structure common to all the cities & the private Prytanæa in time grew out of use. . . The ancient nations built the front of their Temples toward the East & therefore Moses in doing so retained the religion of his ancestors. The placing the fire in the common centre of the Priests Court & of the outward court or court of the people in the Tabernacle & in Solomon's Temple [& the framing the Tabernacle & Temple so as to make it a symbol of the world] is a part also of the religion which the nations received from Noah, for they placed the fire in the middle of the Prytanæa.[305] (Newton, undated-c)

For Newton, the entire heavens were to be

> the true & real Temple of God & therefore that a Prytanæum might deserve the name of his Temple they framed it so as in the fittest manner to represent the whole system of the heavens. A point of religion then which nothing can be more rational.[306]

The Prytanæum embodied universal knowledge,

> So there was one design of the first institution of the true religion to propose to mankind by the frame of the ancient Temples, the study of the frame of the world as the true Temple of the

> great God they worshipped. And thence it was that the Priests anciently were above other men well skilled in the knowledge of the true frame of Nature & accounted it a great part of their Theology. (Newton, undated-c)[307]

The frame of the Prytanæum or Temple embodied the original religion that symbolised the geometric structure of the universe. This geometric structure was the mathematical form of the universe untainted by the corruption of the original religion. In turn, the Temple of Solomon, which replicated the plan of the Tabernacle of Moses, embodied the perfection of the original religion within its structure, which had been inherited from the time of Noah. Babson MS 0434 carefully constructed the Temple through its measurements and its geometry. To understand the frame of the Temple was to understand a great part of the original religion's Theology; the frame of the Temple was the symbol of the exoteric knowledge while the enactment and understanding of the rituals within the Temple lead to the esoteric knowledge of the prophets.

Newton used a range of ancient sources to support his hypothesis: Josephus; Diodorus; Plato; Strabo; Herodotus; Eusebius and many other ancient authors, plus various contemporary authors. Frequently cited are Samuel Bochart's *Geographia sacra;* John Marsham's *Chronicus can aegyptiacus, ebraicus & graecus* and Gerard Vossius' *De theologia gentili.* These three authors created classifications of the pagan traditions to reinforce Christian beliefs. Vossius' *De theologia gentili* was published together with his son, Dionysius Vossius' translation of Maimonides treatise of idolatry, *Mishneh Torah, Hilkhot Avodah Zarah.* Originally published in 1641, it was reprinted eight times by 1700.[308] Newton's copy of *De theologia gentili* showed pages that were "very extensively dog-eared with 112 pages still turned and several similar signs".[309] Kenneth J. Knoespel has demonstrated that Vossius' discussion of the sun's assimilation into religious practice provided a direction for Newton's research. In a section entitled *Quae est de cult corporis caelestis (The use of celestial bodies within religious cults)* Chapters 1–17 Vossius marked out a structure that could have assisted Newton's research. Vossius based his work on a scholarly study of nature in an attempt to reveal the nature of God.[310] He outlined all the different sun cults in ancient religions and considered the way that sun still played a role in Christianity. He considered the ancient observation of the sun's characteristics and verified the sun's velocity mathematically. Although there are similarities between the "mythographic work of Vossius and Newton, the *Origines* does not imitate *De theologia gentili* but simplifies and systematise[s] Vossius' rendition of mythographic material".[311]

In Maimonides' *Laws Concerning Foreign Worship*, he outlined how the earliest man corrupted the true religion by worshiping the stars as objects of veneration. Noah and Abraham were among these who attempted to preserve the original religion. There are certain elements of Newton's concept of the Prytanæa that parallel those of Maimonides. However, the Prytanæa that preserved scientific knowledge and that deified ancestors does not come from Maimonides. In Vossius' commentary on Maimonides, he used Maimonides to develop his own taxonomic analysis of ancient religions, and Newton was closer to Vossius than to Maimonides in some of his concepts.[312]

For Newton, the frame of the world and therefore the Temple was concentric, with the sun or the hearth in the centre. The natural philosophy of the ancients became corrupt along with their religion; the centralized fire was taken to be in the centre of the earth. Thus, the earth replaced the sun in the centre of the universe and this became fully elaborated in the system devised by the second century Egyptian astronomer Ptolemy.[313] The Egyptians were not only the source of the original religion, they were also the source of its demise.

The Prytanæum or Temple as Microcosm of the Macrocosm

The Prytanæum or Temple of Solomon, as the frame of the universe, as the microcosm of the universe, was a widely debated topic in the early seventeenth and eighteenth centuries. In 1604, *Ezechielem Explanationes et Apparatus Vrbis Templi Hierosolymitani* was published; it was to be a collaboration by two Spanish Jesuit priests, Jerome Prado and Juan Bautista Villalpando. However, the early death of Prado left the entire project for Villalpando to complete on his own. *Ezechielem Explanationes* is a commentary on the Book of Ezekiel in three massive volumes. Volume Two contains an elaborate reconstruction of Solomon's Temple. Although there had been reconstructions of Solomon's Temple and commentaries on Ezekiel before the publication of *Ezechielem Explanationes* in 1604, it nevertheless stimulated an avalanche of support, criticism, commentaries and a variation of reconstructions of the Temple of very different opinions and designs. *Ezechielem Explanationes* was an expensive publication and was only made possible by the patronage of Philip II, King of Spain.[314] It was illustrated with elaborate and skilfully executed engravings, and the design of the Temple was fully articulated with plans, elevations and sections. Villalpando had conceived of the Temple as a massive classical edifice, ornately and richly decorated. Both critics and supporters agreed that it was a magnificent design, but many of the commentaries were negative.[315] Newton claimed that "Villalpando, although the best (and) the most eminent commentator on Ezekiel's Temple: yet (he is) out in many things".[316] Newton's comments on the reconstruction of Villalpando are in two manuscripts, Babson MS 0434 and Yahuda MS 14. His comments are a mixture of criticism and support; Newton was highly critical of the architecture of Villalpando's reconstruction, but Newton was supportive of the theoretical justification of the plan (Fig. 4.1).

The basis of Villalpando's reconstruction is that the Temple of Solomon was the microcosm of the universe. Villalpando carefully defined all the measurements of the Temple as being derived from the Sacred Texts, drawing on support from profane texts such as Josephus. He demonstrated that all of the columns of the Temple were in harmonious ratio to each other and to the rest of the building. Villalpando claimed that these harmonic proportions are most apt for a building of divine origins and he implied the existence of a link between the harmonic proportions and the celestial bodies. For Villalpando, the Temple reflected the creation of God and thus needed to incorporate itself into the universal harmony according to the movements of the

Fig. 4.1 Villalpando's plan for the layout of the tribes of Israel's camp around the Tabernacle[317] (Drawn by the author from Villalpando and Prado, 1604, vol. 2, p. 467.)

planets and the fixed stars. To this end, he examined the Tabernacle of Moses, since it prefigured the plan of the Temple, and the camp of the tribes of Israel that surrounded the Tabernacle which is a primitive plan of the Temple precinct. Villalpando first established that the proportion of the atrium that surrounded the immediate temple and the altar was a double square. He then considered the configuration of the camp of the tribes of Israel. The configuration of the camp was highly structured with the Tabernacle placed in the centre, fortified by the four Levites' camps: Moses and Aaron; Caathi; Gerson and Merari. Surrounding them were the twelve tribes of Israel, each tribe camped under a banner that declared its ancient lineage.

Villalpando described the banners that formed the four angles of the square of the precinct of the Tabernacle. In the south-west corner was the tribe of Ephraim and their emblem was a bull and their colour was gold like chrysolite; in the south-east corner was the tribe of Ruben, and their standard was a human face and their colour was red like carnelian; in the north-east corner was the tribe of Judah and their emblem was a lion and their colour was green like emerald. Finally, in the north-west corner was the tribe of Dan whose banner was red and white like jasper, but Villalpando did not clearly state, in this chapter, what the emblem is. He eventually claimed that Dan is like a horned viper, but instead of being represented by a horned viper, many Doctors of the Church and commentators, including Saint Jerome, represented Dan with an eagle. Villalpando demonstrated that the horned viper had the same characteristics as the eagle. The emblem of the flag of Dan is not satisfactorily resolved until the next chapter, where it is made clear that the emblem of Dan is a scorpion. The chrysolite, emerald, carnelian and jasper are four of the twelve gems of the breastplate of the high priest's ceremonial vestment; each one of these twelve stones represented one of the twelve tribes of Israel. The order of the stones on the breastplate is set out in a four by three grid,[318] but only jasper is a corner gem. A calf, a man, a lion and an eagle are the symbols of the Evangelists by Villalpando's placement, which has no Biblical precedent, and with his eagle-like scorpion Villalpando made the Evangelists' emblems the corners of the Tabernacle precinct. Although the eagle as an emblem of Dan was not mentioned by Villalpando after this chapter, later writers including Newton mistakenly acknowledged the role of the Evangelists in the plan of Villalpando's reconstruction.

The distribution and placement of the tribes in the camp was determined by a perfect plan; nothing was left to chance, since it reproduced the plan of the Temple but with its dimensions doubled. It represented the microcosm of the universe – the macrocosm. The four Levite tents that surrounded and fortified the Tabernacle in the plan of the Temple corresponded with the four simple elements of the sub-lunar world, and represented the world of man. These were encircled by the celestial orbits made up of the seven atriums. The orbits are positioned on the plan as Ptolemy assigned them in the *Almagest*[319]: "Thus Saturn is situated between Capricorn and Aquarius; Jupiter in Pisces; Mars in Aries; Venus in Libra; Mercury in Virgo; the Sun in Leo and the Moon in Cancer".[320] Surrounding the seven courts or celestial orbits were the twelve fortifications or bastions of the Temple precinct perimeter. These fortifications corresponded to the twelve tents of the tribes of Israel that where laid out under their banners or standards that declared their ancient lineage: Judah was represented under the symbol of a lion; Ruben under the symbol of the Water-bearer; Ephraim under the symbol of the Bull and Dan under the symbol of the Scorpion (no longer associated with an eagle) and so on so that the tribe's banners were equated to the twelve signs of the zodiac. In the centre was the Temple, "dedicated to the profit of man," that represented the "true Sun" of the super-celestial world of the Church. This true Sun is Christ, the "Sun of Justice" whose light is salvation. This light illuminated the seven planets and the twelve constellations, and the centralized Earth is illuminated by the planet sun that is located in Leo (Fig. 4.2).

The circumference of the heavens is divided into three hundred and sixty degrees due to the movement of the sun that returns in a circuit twenty-four hours around the centralized earth. The diameter of the heavens is one-third of its circumference. The height of the Temple is one hundred and twenty cubits, which coincides with the width of the celestial orbit. The atrium, destined to be a residence of the men, is sixty cubits in height, half the circumference of the heavens – i.e. man dwells under the heaven of heavens. This perfect plan represented the three worlds of the microcosm and macrocosm. In the centre was the super-celestial world of God, this is surrounded by the world of man, and then the celestial world of the seven planets and the fixed stars encircling the Earth – a perfect vision of a geo-centric universe.

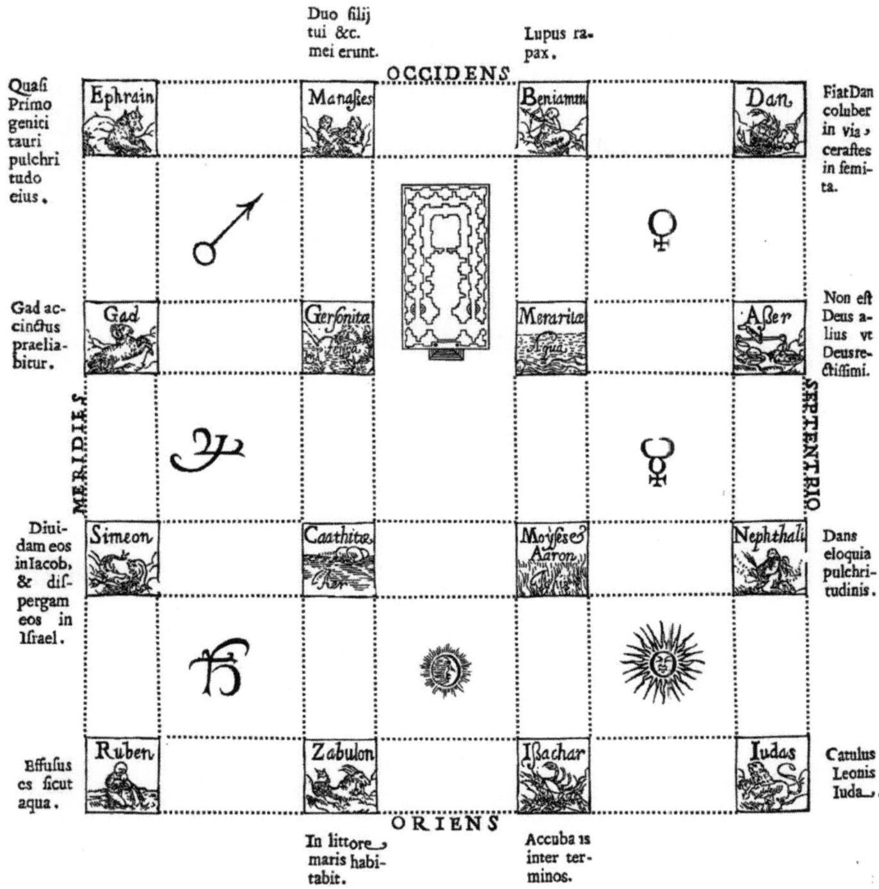

Fig. 4.2 Villalpando's astrological arrangement for the plan of Solomon's temple[321] (Drawn by the author from Villalpando and Prado, 1604, vol. 2, p. 470.)

Villalpando fully endorsed the anthropomorphic theories of Vitruvius. He perceived that the humanity assumed by God is reflected in the measurements and geometry of the Temple, which prefigured the perfection of the mystical body of the Church. The measurements and the proportions of the Temple are reflected in man. The measurements of the Tabernacle equate to the ages of man's active military service; the age of twenty is the age to enlist, twenty-five the age of perfect strength and the age of fifty is the time of weakening strength. This emphasized the Tabernacle precinct as the camp of the twelve tribes of Israel, whilst the proportions of the temple equate to the proportions of man. Man has a height of six feet, this measurement agrees with his arms extended; but if the arms are doubled in front of the chest, so that the end of the longest finger of the right hand touches the end of the middle finger of the left hand, then the width of man will be one and a half cubits or three (Roman) feet. The colonnades of the Temple have eight inter-columns, which coincide with the height of the head of man from the chin to the upper part and are divided into three promenades or galleries that correspond to the barrel of the chest and with the arms. These colonnades correspond to the proportion of 1:2, not only a double square, but also the harmonic ratio of an eighth – an octave. Here, Villalpando portrayed Christ taking the appearance of man as the cosmological man, which emphasised the microcosm-macrocosm analogy (Fig. 4.3).

The gridded floor plan of Villalpando's reconstruction that corresponded to the plan that represented the microcosm of the universe was crowded with colonnades and incorporated 1,500 columns. The Temple precinct was 500 × 500 cubits and the

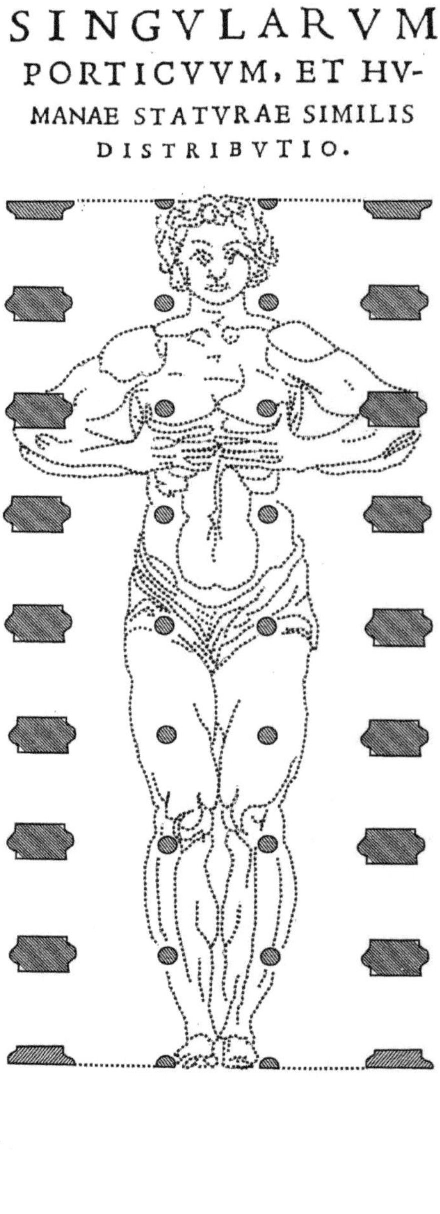

Fig. 4.3 Villalpando: A single colonnade and the resemblances to the division of the human stature[322] (Drawn by the author from Villalpando and Prado, 1604, vol. 2, p. 472.)

exterior boundary 800 × 800 cubits. Its height, including the foundation, was a massive 420 cubits. Every part or element was in a harmonious ratio to the entire building. For Villalpando, this was the greatest building ever built and no building could ever surpass it. His was the first full-scale reconstruction of the divine archetype and this reconstruction inspired other commentaries and other reconstructions of Solomon's Temple.

There were six main points of debate identified by Villalpando's critics that were stimulated by *Ezechielem Explanationes*. First, the Divine origins of the Temple were questioned: was God the architect of the Temple? Second, Villalpando's reconstruction had no historic basis. It was far too elaborate for the tenth century BC and it would not have been built in the classical style. Third, the Temple's architecture was not the pinnacle of architecture and the design would be surpassed by subsequent designs; in particular, Herod's Temple was larger and grander than Solomon's Temple. Fourth, the interpretation of the Biblical measurements of a cubit by Villalpando was wrong and the result of this was that Villalpando's plan exceeded the site of the Temple at Mount Morion. Fifth, the lack of Jewish sources in Villalpando's work, such as the *Talmud*, *Middoth* and Maimonides' description of the Temple in Book Eight of *The Code of Maimonides (Mishneh Tornh)*,[323] gave a limited view of the Temple. Last, Ezekiel's vision of the Temple was not the same as the Temple of Solomon. It was the last two points regarding the sources of the Temple that generated the most criticism and, in turn, this generated a large number of reconstructions from various sources in response.

One criticism emerged before the publication of *Ezechielem Explanationes* from a fellow Jesuit, Benito Arias Montano. Montano criticised Villalpando for his use of Ezekiel's vision, and he claimed that this was not the same as Solomon's Temple. Montano based his reconstruction primarily upon the Book of Kings (see Fig. 4.4).

Claude Perrault, architect of the Louvre, illustrated *The Code of Maimonides, the Mishneh Torah* which had been translated into Latin by Louis Compiègne de Veil in 1678 (Fig. 4.5).[325] In the Preface of de Veil's translation he expressed surprise that Villalpando had spend so much time and effort on an image that did not portray the historical truth. He claimed that Villalpando's main aim was to prove that the Temple conformed to Vitruvian norms and that the Greeks and Roman had learned the art of building through studying the ancient buildings.[326] Constantin L'Empereur (Fig. 4.6), John Lightfoot and Louis Cappel were interested in historic reality and they considered that Villalpando's reconstruction had failed because it had not considered the Jewish tradition. They, in turn, were criticised by Bernard Lamy because they had failed to realise that the Jewish writers on whom they had based their reconstructions were inexperienced in building and were incapable of giving a competent account of the Temple.[327] The plans that were derived directly from the Jewish sources were notably different from those used by Villalpando and Newton; importantly, the buildings of the Temple were not symmetrically placed within the Temple precinct. These plans would not have been suitable for either Villalpando's image of the Temple as the microcosm of the universe or Newton's frame of the world. In all, the criticism of Villalpando, which was extensive,[328] the Temple as the microcosm was not criticised and appeared to have been a generally accepted concept.

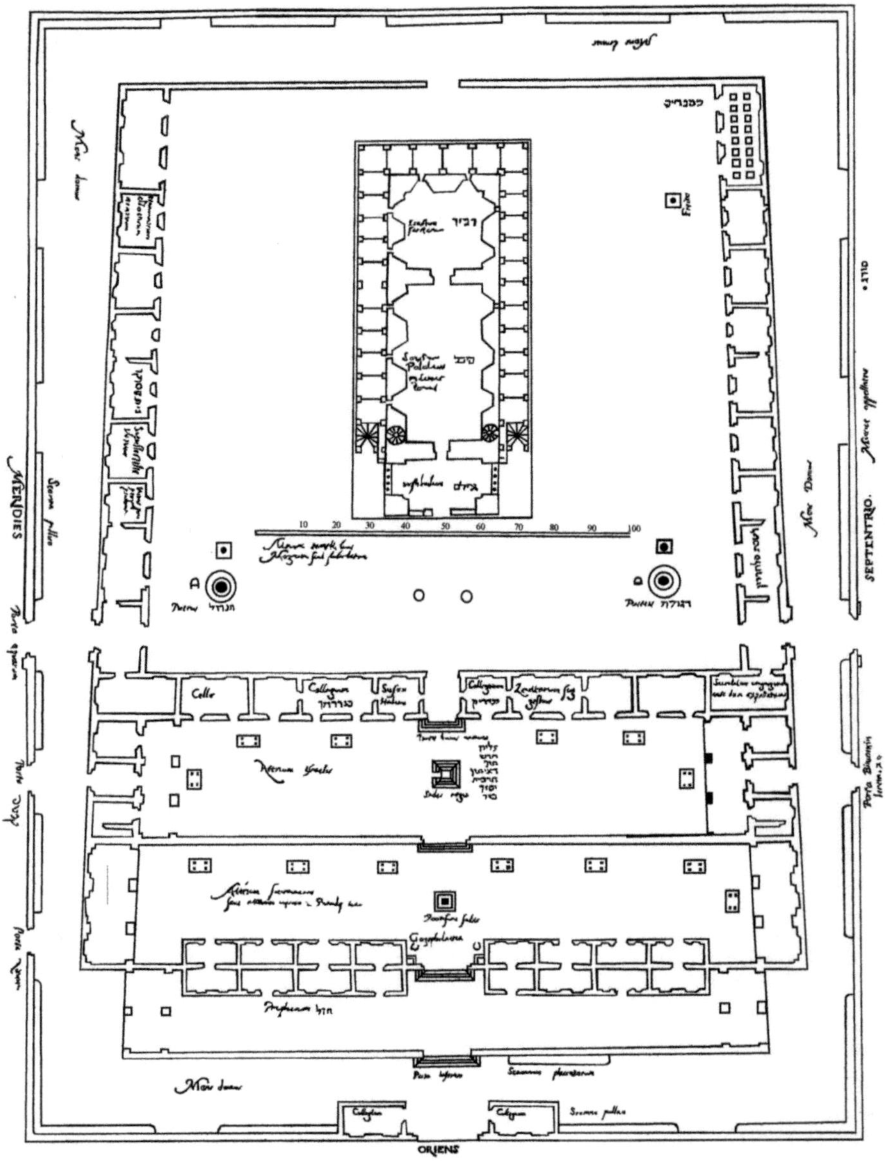

Fig. 4.4 Montano's reconstruction of the Temple of Solomon, from his "Exemplar" in volume eight of the Antwerp Polyglot[324] (Drawn by the author from Zur Shalev (2003), p. 64, with kind permission.)

As stated previously, Newton's comments on Villalpando were a mixture of both support and criticism. Like Villalpando, Newton strongly believed that Ezekiel's vision of the Temple was the same plan as Solomon's Temple. Also, like Villalpando, Newton reconstructed the structure of the Temple to reveal it to be mathematically perfect. However, his floor plan and description of the Temple were remarkably

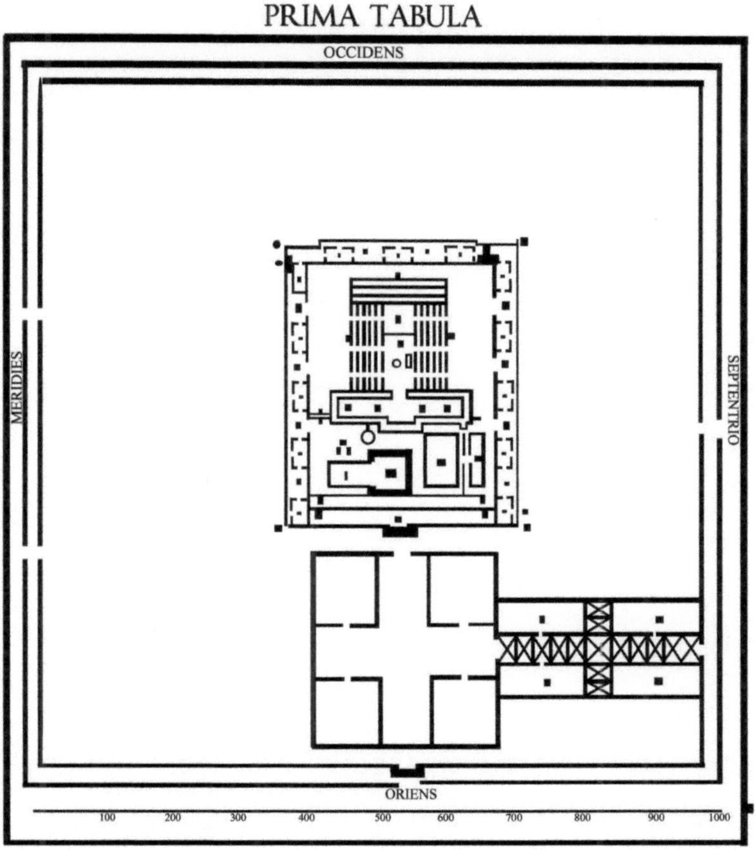

Fig. 4.5 Claude Perrault's floor plan of the Temple from Louis Compiègne de Veil's translation, from Hebrew into Latin, of *The Code of Maimonides, the Mishneh Torah* originally published in 1678[329] (Drawn by the author from Louis Compiegne de Veil, 1683, unpaginated.)

different from Villalpando's, despite being derived from the same source, that of Ezekiel. He believed that Villalpando's errors in his design had primarily originated from his failure to take advantage of Jewish sources and from his misinterpretation of the Latin texts.[330] Newton pointed to the Latin text that Villalpando had used as being sometimes different in its translation to the Hebrew texts. For instance, in the Latin version in Ezekiel 42:3, Villalpando translated "colonnades united" to be a triple colonnade but according to Newton in the Hebrew text it translated to "colonnade against colonnade three times" indicating three storeys.[331]

Villalpando created his grid plan of the Temple precinct from what Newton considered an "incorrect translation"; Newton also said that his plan "has no support and is lacking in reason".[334] Villalpando interpreted Ezekiel 40:19–20 as meaning the length of the atrium from the south to the north to be the distance between the gates, a hundred cubits, and this divided the area of the precinct into small atriums or ante

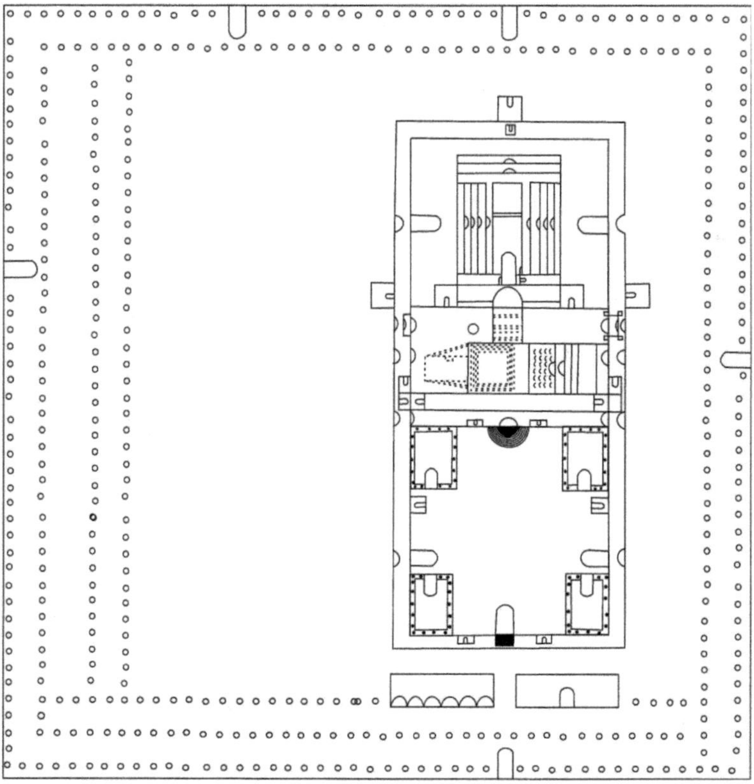

Fig. 4.6 Constantin L'Empereur's floor plan of the Temple from Guglielmus Surenbusius; Mishnah sive Legum Mischnicarum liber qui inscribitur Ordo Sacrorum..., 1702[332] (Drawn by the author from James Stevens Curl, 1991, p. 89, with kind permission of Professor James Stevens Curl.)

rooms, one larger one that formed the temple atrium and seven exterior to it (see Fig. 4.7). These ante rooms are divided from each other by triple colonnades of fifty cubits in width. Newton pointed out that not only are these ante rooms not mentioned in Ezekiel, but that the thirty chambers that are next to the sides of the gate and that are expressly mentioned by Ezekiel are absent. However, Villalpando recognized that in his reconstruction these chambers were impossible if the spaces of the gates were not counted, but this went against the text of Ezekiel. In addition, Newton also claimed that Villalpando's grid plan cannot be accepted

> unless we want to move away from the proportion of Moses' atrium that surrounds the immediate temple and the altar, which was established by Villalpando himself as being a length over double its width.[335]

These criticisms, based on Villalpando's interpretation of the Biblical texts, challenge the entire basis of his reconstruction. The triple colonnades that Newton claimed were a mistranslation were important to Villalpando's plan. First, they portrayed man/Christ as the cosmological man, emphasizing the microcosm-

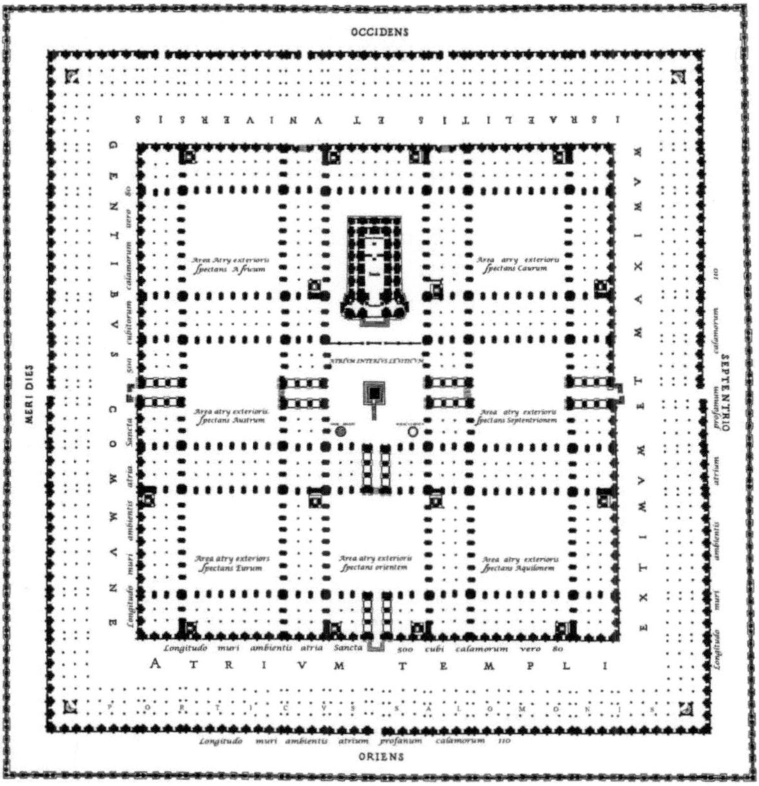

Fig. 4.7 Villalpando's reconstruction of the temple[333] (Drawn by the author from Villalpando and Prado, 1604, vol. 2, unpaginated.)

macrocosm analogy. Second, they divided the gridded plan into the seven small ante rooms and the temple atrium, which Newton considered to be "lacking in reason", and their creation, went against the proportions of the Temple atrium that Villalpando had himself established. These triple-colonnaded atriums not only formed a considerable part of Villalpando's reconstruction, they are also significant for the plan of the astrological plan of the macrocosm. Their removal from his plan changed his reconstruction to an unrecognizable degree. Furthermore, Newton referred to Villalpando's reconstruction as a "fantasy".[336] In fact, there was not much about Villalpando's reconstruction that appealed to Newton. This begs the question as to why Newton did consider Villalpando as "the best (and) the most eminent commentator on Ezekiel's Temple".

In Yahuda MS 14[337] Newton accepted that this plan prefigured the plan of the Temple and the proportions of the temple, which were double than that of the Tabernacle as proven in detail by Villalpando, but which Villalpando himself seemed to have forgotten when constructing his own floor plan. In addition, Newton agreed that the perfect architectural harmony of the Temple represented in microcosm the perfect harmony of the macrocosm. However, Newton misread Villalpando's geo-centric plan of the microcosm-macrocosm; he considered it to be a heliocentric system. Newton accepted Villalpando's symmetrical layout of the camp around the Tabernacle, but he also misread the heraldry of the tribe's standards. Newton took the symbol of Dan to be the eagle, when it was in fact a scorpion, which made the four corner standards of the tribes of Israel the symbols of the Evangelists (Newton, 1999). For both Newton and Villalpando the Temple of Solomon was the divine plan of God and represented the microcosm of the universe.

Stonehenge as Prytanæum

Newton's concept of the Prytanæum did have ancient precedents, but in the seventeenth century the image of the Temple had been brought into a public debate by Villalpando's reconstruction, which also promulgated the concept of the Temple as a microcosm of the macrocosm.

Newton claimed that the Prytanæum was universal in the ancient religions and which was evident in England, Denmark, Medes and Persia, Ireland and throughout the ancient world.

> In England near Salisbury there is a piece of antiquity called Stonehenge which seems to be an ancient Prytanæum. For it is an area compassed circularly with two rows of very great stones with passages on all sides for people to go in and out at. Tis said that there are some pieces of antiquity of the same form & structure in Denmark. For its to be conceived that the Vestal Temples of all nations as well as of the Medes & Persians were at first nothing more then open round areas with a fire in the middle, till towns & cities united under common councils & built them more sumptuously. In Ireland one of these fires was conserved till of late years by the Moncks of Kildare under the name of Briget's fire & the Cænobium was called the house of fire.[338]

A lot has been made of the fact that Newton described Stonehenge as one of these ancient Prytanæum.[339] However, he only mentioned it once and the concept of Stonehenge as an ancient Temple had been well established by Inigo Jones in *The Most Notable Antiquity of Great Britain Vulgarly called Stone-Heng on Salisbury Plain Restored*[340] published in 1655 and John Webb in *A Vindication of Stone-Heng Restored*[341] published in 1665.

Jones surveyed Stonehenge in 1620 at the request of King James,[342] and he continued to work on his "architectonical scheme" up to his death in 1652. *Stone-Heng Restored* was edited by Webb and published three years after Jones' death. Jones' original notes for the book are missing, thus it is impossible to say how much of the work is Jones' and how much is Webb's.[343] Nevertheless, the work is significant in that it reveals both Jones' and Webb's attitude towards the connection between Classical composition and the celestial symbolism of the ancient style of Temple and the Temple of Solomon.

Jones perceived Stonehenge to be a Roman Temple constructed of Tuscan columns and dedicated to Cælus, god of the sky. He considered that

> the Romans for so notable a structure as Stone-Heng, made choice of the Tuscan rather than any other Order, not only as best agreeing with the rude, plain, simple nature of those they intended to instruct, and use for which erected; but also, because presuming to challenge a certain kind of propriety therein, they might take occasion thereby, to magnify to those then living the virtue of their ancestors for so noble an invention, and make themselves the more renowned to posterity, for erecting thereof, so well ordered a building.

Besides, the Order is not only Roman, but also the scheme (consisting of four equilateral triangles, inscribed within the circumference of a circle) by which the work Stone-Heng formed was an architectonical scheme used by the Romans.[344]

Jones constructed a composition for the Temple that had little real relation to the structure; it consisted of four equilateral triangles within a circle, the overlaying equilateral triangles formed a four-sided tetragon and the six "columns" in the centre formed a hexagon (see Fig. 4.8). He related this architectonical scheme to Book Five, Chap. VI of Vitruvius' *De architectura,* where there is a plan of a theatre. Jones quoted Vitruvius as saying that at the base of the theatre (Fig. 4.9):

> let four triangles be inscribed of equal sides and intervals, which may touch the extreme part of the circumference; by which figures also, astrologers from the musical harmony of the stars ground their reasonings as concerning the description of the twelve celestial signs.[346]

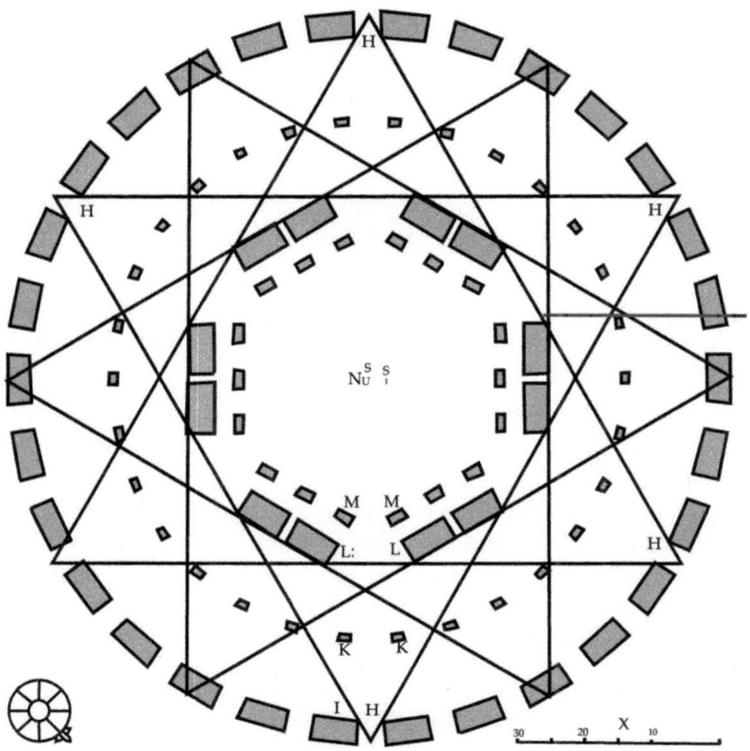

Fig. 4.8 Inigo Jones' architectonical scheme for Stonehenge[345] (Drawn by the author from Inigo Jones, 1655, pp. 58 and 59.)

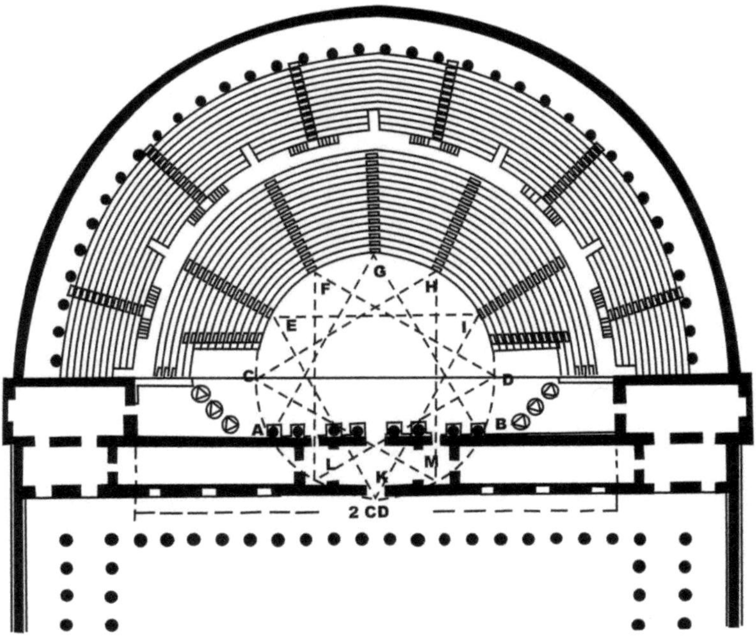

Fig. 4.9 Vitruvius' plan of the theatre[347] (Drawn by the author from Vitruvius, 1960, p. 147 with kind permission.)

The composition within the circle of Jones' architectonical scheme consists of triangles, tetragons and a hexagon. Jones quoted French humanist and commentator on Vitruvius, Guillaume Philandrier, as saying "The astrologers make use of three sorts of figures; the triangle, tetragon and hexagon".[348] For Jones,

> Now this Antiquity consisting of several stones, orderly disposed into one entire work, in imitation, as it were, of those several stars which appearing to us in the heavens in the form of a circle, are called the celestial crown; and the wholly designed by those schemes wherewith astrologers use to describe celestial bodies.[349]

In the Temple of Stonehenge, Cælus, the god of the sky was worshipped through sacrifices which were performed around a fire. The significance of the fire is reflected in the structure of the columns or upright stones, "all the upright stones in this Antiquity are pyramidal like flames, in imitation of those ethereal fires, wherewith the heaven is adorned".[350] Jones' description of this ancient "Temple" Stonehenge parallels Newton's concept of the Prytanæum. Furthermore, Jones related his architectonical scheme to the Temple of Jerusalem. This architectonical scheme that the astrologers used to describe the celestial bodies,

> being all jointly made use of by the architect for conformation of this sacred structure, it is not impossible Stonehenge was so composed, because dedicated to Cælum. Yea further, (if lawful to compare an idolatrous place with so divine a work) was not the Temple of Jerusalem adorned with the figures of Cherabim, that thereby the Nations of the Earth might know it was the habitation of the living God? And, why not in this manner this temple composed by astrological figures, that after Ages might apprehend, it was anciently consecrated to Cælus or Cælum Heaven?[351]

Although Newton developed these ideas, the concept of ancient Temples, including Solomon's Temple, representing the microcosm was well established in the seventeenth century.

Stukeley published *Stonehenge a Temple Restor'd to the British Druids*[352] in 1740. His researches stem back to 1721–1724 and there are existing manuscripts that contain his field work from this time.[353] He compared Stonehenge to the fabric of Solomon's Temple and found that it was built using the cubit. This cubit of the Druids was the same as the Egyptian and Hebrew measurements of the Bible.[354] Stukeley examined the measurements of Stonehenge and the Druidic cubit; his examination could have been stimulated by Newton's study on the ancient cubit entitled *A Dissertation upon the Sacred Cubit of the Jews,*[355] although this was published posthumously in 1737. The original paper is entitled "De magnitudine cubiti scari" and is part of a much larger manuscript; it is loosely dated late 1670s–1690s, and it is possible that Stukeley was aware of this work. However, Stukeley, who was always keen to align himself with Newton, did not mention him having any interest in Stonehenge in his biography of Newton, nor did he mention discussing ancient measurements with him. It does appear likely that Newton was only interested in Stonehenge as an example of the ancient Prytanæum of England and nothing more.

Villalpando, Jones, Newton and Stukeley applied the norms of Vitruvius to the ancient Temples. For Villalpando, Jones and Stukeley the norms of Vitruvius were derived from pure natural reason. Both Jones and Stukeley defined Stonehenge in classical terms and believed that these norms of architecture were derived from nature. Villalpando clearly distinguished sacred architecture from the profane architecture of Vitruvius. He claimed that "Sacred architecture constitutes the origin of architecture, and the profane one is like a copy, or better still, as a shadow of sacred architecture".[356] The purpose of Vitruvius, who Villalpando described as "the pioneer of our architects," was to equip the architect with the norms of architecture. But Villalpando's purpose was to examine the origins of architecture and to extract the norms of architecture that were derived from God's plan and promulgated by the Scared Scriptures, and this natural order was followed by Vitruvius in his *Ten Books on Architecture*. Villalpando's reconstruction envisaged the Temple to be a building that encapsulated the entire formal grammar of classical architecture, which begins with the harmonic ratios. In Babson MS 0434, Newton only mentioned Vitruvius by name once, but he considered how the Temple was built to the "proportion of the architecture"[357] and these proportions parallel Vitruvius' norms.

Maimonides' Floor Plan of the Temple

The floor plans in Babson MS 0434 and the other plan in *Chronology* are concentric, with the altar of the Temple in the centre. Newton criticised Villalpando for not taking advantage of the Jewish sources and for misinterpreting the ones that he did use.[358] He also criticised Louis Cappel and Arias Montano for departing from rabbinical material.[359] However, he did not elaborate on how they departed from the material.

Newton was very selective on the Jewish sources he used and Newton's most notable departure from the Jewish sources was in the plan of the Temple, particularly, as laid out by Maimonides in *The Code of Maimonides: Book Eight: The Book of Temple Service*.

Maimonides' description was about the Temple of Jerusalem, not necessarily the Temple of Solomon. He claimed:

> The Temple building erected by Solomon is clearly described in the Book of Kings. Furthermore, the building to be erected in the future, even though it is discussed in the Book of Ezekiel, is not fully described and defined therein. Therefore, those who build the second Temple in the days of Ezra followed the pattern of Solomon's Temple and adapted some of the particulars described in Ezekiel.[360]

He all the sources for his reconstruction and although there are differences in the detail and the grandeur of the buildings of the second Temple precinct, it was built on the same foundations as Solomon's.

There are many differences in the plan of Maimonides' Temple to that of Newton's, such as the numbers of chambers and the heights of parts of the buildings. Nevertheless, there are two significant differences between the plans which, in both cases, are attributed to Solomon. These are the design of the altar and the layout of the floor plan of the Temple precinct.

Maimonides claimed that the dimensions of the altar were very precise and that the design was handed down from one generation to another since the time of Solomon.[361] It was thirty-two cubits in breadth and width, and ten cubits in height. However, some of the cubits in height were sacred cubits (that equal six palms) and others were the vulgar cubit (that equal five palms). This mixture of sacred and vulgar cubits made the altar fifty-eight palms in height.[362] For Newton, the altar of Solomon was twenty sacred cubits in breadth and width and the height was ten sacred cubits.[363] He claimed that the altar remained the same for the second temple and even until after the time of Alexander the Great,

> but later, upon not understanding the mathematical expression 'to carry the length to the width,' the words of Ezekiel were interpreted erroneously as if the length and the width of twelve cubits had itself to be measured from the centre of the altar. And thus, adding twelve cubits to the correct dimensions, they built an altar of thirty-two cubits of length and width in the base.[364]

However, Newton did not mention Maimonides as being one of the "laters". Apart from the size difference there was one important feature that Newton did not mention; a ramp for the priest to be able to serve at the altar. Maimonides' ramp was a massive thirty-two cubits in length, sixteen in breath and rose to a height of nine cubits; it was situated on the southern side of the altar.[365] Ten cubits is close to five metres and according to the Book of Exodus "Neither shalt thou go up steps unto Mine altar".[366] Newton did not indicate how the priest would have been able to serve at the altar, yet this was a significant feature in the design and ritual of the Temple.

The layout of the floor plan of the Temple precinct of Maimonides was significantly different from that of Newton. Like Newton's plan, the Temple precinct was square and the walls were five hundred cubits in length on each side. However, the

Great Court that included the Temple and the altar was one hundred and eight-seven cubits from east to west; and one hundred and thirty-five cubits from south to north. Maimonides placed the Great Court

> not exactly in the centre of the Temple Mount. It was further from the southern wall of the Temple Mount than from any other side, and closer to the western wall than to the other side. Between it and the northern wall there was a greater distance than between it and the western wall, and between it and the eastern wall a greater distance than between it and the northern wall.[367]

Clearly, Maimonides' floor plan, with the ramp of the altar on the southern side, would have ruined the symmetry that was paramount to Newton's floor plan.

Notably, Maimonides' plan could not be considered a motif of the microcosm since it was not concentric and had no "hearth" in the centre. The concept of the "hearth" in the centre of the concentric Temple was an important concept for Newton for several reasons. The concentric plan of the Temple with the priests sacrificing at the centralised altar recalled the original rituals of Noah. It represented the sacred sacrificial fire of the true ancient religion which had knowledge of the universe. Additionally, the sacred architecture of the Temple represented knowledge of the universe; within its design was encoded the cosmic harmonies which were applicable to the heliocentric system. Finally, the Temple plan held the esoteric knowledge of God's universe; thus, with Newton's exoteric knowledge of the universe through the *Principia,* the metaphysical and the physical were brought together in number, weight and measure.[368] In the Book of Wisdom, God has ordered "all things in measure, and number, and weight".[369] For Newton, the metaphysical could be understood through the rituals and prophecies of the Temple which were expressed or enacted within the harmonic and geometrical architecture of the Temple, and were applicable to the heliocentric system. The physical could be understood through the geometry of the heliocentric system which was expressed in the *Principia.* Maimonides' off-centre model of the Temple did not fit this image, but Newton did not dismiss the model, he just ignored it.

Conclusion

The concepts that were begun in the *Origines* on the corruption of the church were never fully refined by Newton, even though he returned to the topic repeatedly throughout his life. However, he strongly linked the loss of ancient knowledge of natural philosophy with the corruption of the church. This ancient knowledge had been preserved in the rituals of the Prytanæa, whose sacrificial altar, the hearth, the centre of their worship represented the structure of the heliocentric solar system. Their temples, the Prytanæum, embodied the truth of natural philosophy and their purpose was to represent God's cosmos to the people. For Newton, the plan of Solomon's Temple was an example of one of the Prytanæum, the plan had come from God through Moses, and had been preserved by the prophets in the Biblical texts.

In the mid-1680s, the Temple became the hieroglyph for the universe and the sacrificial altar, the hearth as the sun became dominant in Newton's unpublished papers; this coincides with his writing of the *Principia*. In the Sacred architecture of the Temple was encoded the cosmic harmonies of the planetary motions, the ancient knowledge of the heliocentric frame of the universe – in short the Temple held the esoteric knowledge of the universe.

Measurements and the act of measuring are extremely important elements of the Biblical description of both the Temple and the Apocalypse. Ezekiel and John the Divine are guided around the Temple and the New Jerusalem respectively by an Angel who measures both the buildings. The measurements in both cases are the most dominant feature of the building and in Newton's reconstruction of the Temple in Babson MS 0434 he emphasised the importance of the measurements by demonstrating how they fit together. The measurements are in cubits. Cubit means elbow, a vague description which lacks any precision. To make matter worse there were many types of cubit: Roman cubit; Greek cubit; Arabian cubit; Simple Egyptian cubit; Royal Egyptian cubit; Memphis cubit; Babylonian cubit etc. In the Temple, there is a distinction between the sacred cubit and the vulgar cubit. The Biblical measurements could provide the proportions of the Temple, but without knowing the size of the cubit in modern measurements it was not possible to know the correct scale of the Temple. Newton executed a meticulous study of the cubit to further understand the dimensions of the Temple.

Chapter 5
The Temple Measurements and the Sacred Cubit

Overview of Ancient Measurements

In 2 Chronicles 3:2, Solomon instructed that the Temple be built in cubits "after the first measure". This implies an ancient standard and the cubit was one of the most widely used measurements in the ancient world. It was considered to be a natural measurement. Deuteronomy 3:11 describes the bedstead of Og, the King of Bashan: "nine cubits was the length thereof, and four cubits the breadth of it, after the cubit of a man". To measure the bed with a cubit was to measure it with the length of the forearm from the elbow joint to the end of the middle finger. The body as a measurement was a simple and ever handy measuring stick. This simple measurement is inscribed in Egyptian hieroglyphs. The hieroglyph for a cubit is the image of the forearm,[370] and the cubit can be divided into smaller parts; there are also measurements of the body, of palms and of digits.

Methods of standardising these measurements can be seen in the archaeological record. In the Piraeus Archaeological Museum in Greece a fourth century BC stone engraved with standard measurements has been preserved and it gives the measurement for: an orgyiae, which is the arms fully outstretched; a cubit (43.7 cm); an open palm (24.2 cm) and the foot (32.2 cm) (see Fig. 5.1). There is another Greek measuring standard in the Ashmolean Museum, Oxford. It is a fifth century BC carved relief and the remaining fragment shows the outstretched arms of a man to measure the orgyiae, and the imprint of the foot.[372] The orgyiae is seven times the foot in this relief,[373] which is larger than the idealised human span of six feet that was later made famous by Vitruvius. Also, the Rhind Mathematical Papyrus,[374] c1550 BC, held at the British Museum records the length of the Egyptian Royal cubit as 20.6–20.7 in. or 52.4–52.7 cm. The rod of Haremhab, c1333–1306 BC, records the Egyptian Royal cubit to be 20.4 in. or 51.8 cm.[375] While an Egyptian measuring stick c1069–715 BC held at the Metropolitan Museum, New York measures a unit of 27.5 in. or 69.8 cm.[376] There appears to be no consistency in the standards of these ancient measurements and from the various archaeological sources that have survived it would appear that the "standards" were very localised or at least they were not standardised with any precision. However, Newton did not have the benefit of this evidence. He had Biblical and ancient literary sources, and the only other source

T. Morrison, *Isaac Newton's Temple of Solomon and his Reconstruction of Sacred Architecture*, DOI 10.1007/978-3-0348-0046-4_5, © Springer Basel AG 2011

Fig. 5.1 Greek fourth century BC stone engraved with standard measurements[371] (Drawn by the author at Piraeus Archaeological Museum in Greece)

were the measurements of the Great Pyramid by John Greaves (1602–1652), Professor of Geometry at Gresham College.

There is no Hebrew standard of measurements that has been recovered and all the attempts at finding the length of the Hebrew vulgar and sacred cubit have been through equivalent Roman, Greek, Arabian, Egyptian, Memphis and Babylonian measurements. The Biblical and ancient sources are as inconsistent as the archaeological evidence. Herodotus writing on Babylon claimed that "the royal cubit is three inches longer than the ordinary cubit".[377] Presumably the royal cubit is the Babylonian cubit since Herodotus is in Babylon, although it is impossible to be sure of that as he does not establish what the ordinary cubit is. Sabbath day's journey for the Jews was to be not longer than two thousand cubits.[378] According to Acts 1:12, a Sabbath day's journey was the distance from Jerusalem to the Mount of Olives following the ascension of Christ. But the distance between Jerusalem to the Mount of Olives is described by Josephus as being five Roman furlongs in *Jewish Antiquities*[379] and six Roman furlongs in *the Jewish Wars.*[380] In 1 Kings 7 and 2 Chronicles 4, the Bronze Sea of the Temple of Solomon is described and the measurements are in cubits and its capacity is in baths. In 1 Kings. the sea was 10 cubits in diameter, 5 cubits in height and it was 30 in circumference,[381] and it held 2,000 baths.[382] In 2 Chronicles, the sea was 10 cubits in diameter, 5 cubits in height and it was 30 cubits in circumference,[383] and it held 3,000 baths.[384] These and many other inconsistencies in the ancient texts make any direct method of converting the measurement from one to the other impossible.

John Greaves and Egyptian Measurements

John Greaves (1602–1652), Savilian Professor of Astronomy at the University of Oxford, was a mathematician-Orientalist with a command "of ancient and modern astronomical and geographical literature of Latin, Greek, Hebrew, Arabic and Persian authors".[385] He conducted a survey of the Pyramids of Giza which resulted in the publication of *Pyramidographia* in 1646.[386] Greaves described the physical attributes of the Great Pyramid and was the first to execute an architectural section of the Pyramid. He also attempted to establish its chronology and the history of its construction. Using Biblical prophecies and the epochal dates of the Olympiads, on the authority of Diodorus and Herodotus, Greaves concluded that the Great Pyramid was built at some time between Moses and the destruction of the Temple.[387] Like Newton he shortened Egyptian history.

In a time when Hermetic philosophy was at its zenith, Greaves' aim in studying the Pyramids was not to search for Egyptian esoteric wisdom – it appears that he was uninterested in Egyptian mysticism – but to search for the origins of modern measurement. To achieve many of the measurements of the interior of the Pyramid he had to crawl through dark tunnels and then meticulously take the measurements. Before the Greaves' study the measurement of this dark and mysterious interior had been only made by approximation.[388] Greaves bought mathematic precision and a passion into the study of metrology.

In 1647, he published *A discourse of the Romane foot and denarius, from whence the measures and weights used by the ancients may be deduced*.[389] He claimed "that the foot was the most received, and usual measure among the Romans, as the cubit among the Jews, is not controverted by any".[390] He used a wide range of ancient and modern sources in his study including Polybius, Suetonius, Pertius Vicentiniss, Philandrier, Vitruvius, Donatus, Villalpando and many others. Greaves also examined archaeological evidence in Rome such as the marks on columns and pavement stones at the Pantheon, Via Appia and the Roman brass feet, but he found them disappointing.[391] He constructed a table of all of the variations of the Roman foot and a comparative table in English feet, and concluded that the Roman foot was 0.967 of an English foot[392] or 11.6 in. Greaves does not attempt to calculate the Hebrew cubit; his measurements in English feet were used to calculate the Royal cubits, Memphis cubits and the Egyptian cubits.

Newton's Interpretation of the Hebrew Cubit

In Ezekiel's vision of the Temple, two types of cubits were described which Newton distinguished as the sacred cubit and the vulgar cubit. The description of the cubits in Ezekiel is very confusing as he claimed, "The cubit is a cubit and a palm breath",[393] leaving the distinction between the two cubits ambiguous. One of the essential elements of Babson 0434 is measurements. Everything is carefully measured, those measurements are thoroughly scrutinised and the structure of the building is considered from these measurements and classical architectural proportions. Newton noted the ambiguity in Ezekiel's description of the cubit[394] and notes, "Jerome was reading a corrupted version which he took to serve as an example by the use of the reed being six cubits and a palm and the Latin ones followed him ordinarily".[395] In fact, the text claimed that a reed was six cubits in length and each of these cubits were one cubit plus a palm in length. In other words, the sacred cubit was one palm greater than the vulgar cubit. Newton claimed that a vulgar cubit was five palms and a sacred cubit was six palms.[396] But Newton does not extend the discussion of what would seem to be a salient point in Babson MS 0434. However, he made an extensive study of the cubit which was posthumously printed as "A Dissertation upon the Sacred Cubit of the Jews and the Cubits of the Several Nations".[397] Here Newton examined the measurements taken by Greaves, which he took to determine the Royal cubits, Memphis cubits and the Egyptian cubits.[398] From Greaves' calculations of the ancient cubits, Newton proceeded to calculate the measurement of the Jewish sacred cubit, which was essential for understanding the Temple structure.

Newton's "Dissertation" begins: "To the description of the Temple belongs the knowledge of the sacred cubit; to the understanding of which, the knowledge of the cubits of the different nations will be conducive".[399] Newton used Greaves' measurements of the Great Pyramid and systematically compared them with measurements given by ancient sources such as Herodotus, Vitruvius, Strabo, Josephus, Hesychius of Alexandria, Lucius Iunius Moderatus Columella, Philandrier, Gnaeus Julius Agricola, Publius Clodius Thrasea Paetus, *the Talmud* and more contemporary writers such as Willebrord Snellius, Samuel Purchas and Juan Bautista Villalpando. Greaves quoted from Arabic sources, such as Ibn Abd Alhokm (321–405), who Newton also cited.[400] However, the sources may extend beyond those that Newton directly references. José Faur claimed that Newton alluded to a commentary of R Obadiah of Bertinoro (second half of the fifteenth century) in the "Dissertation"[401] and that he translated "the Aramaic version of Ruth 1:6 in conjunction with the distance a Jew is permitted to travel by foot on the Sabbath".[402]

First, Newton examined the Roman and Greek cubit and foot, measuring them in palms and digits, together with the Greek orgyiae, because these measurements were defined by the ancient authors. To estimate their value, Newton approached the problem of the variations in the measurements of the ancient authors by assessing each one of their limits and then comparing them to each other. Throughout the "Dissertation", various ancient measurements are surveyed and calculated.

Newton reasoned that the builders of the Great Pyramid would have used a uniform unit of measurement in their design – a cubit of Memphis. In his calculations he claimed that one Greek orgyiae is equal to four Memphis cubits. However, Ibn Abd Alhokm had claimed that the side of the Great Pyramid was one hundred Royal cubits. Newton argued that "it is probable, that the Egyptians learned, from the orgyiae of the Greeks, their measure of four cubits of Memphis, and gave it the name of the Royal Cubit".[403] From this point, Newton examines the ancient measurements with Greaves' measurements (see Table 5.1).

Newton pointed to the difference between these measurements of the table as being a mere seventh of any inch, "an error of no importance if we consider the much

Table 5.1 Measurement of the Great Pyramid as given by Greaves and the ancient source

Feature measured	Greaves' measurements (in English feet)	Ancient measurements	Newton's ancient measurement (in English feet)
Length of the base of the Great Pyramid	693	100 Royal cubits = 400 Memphis Cubits	400 × 1.732 = 692.8
The square passageway entrance width and height	3.463	2 Memphis cubits	2 × 1.732 = 3.464
The length of the chamber	34.38	20 Egyptian cubits	20 × 1.719 = 34.38
The width of the chamber	17.19	10 Egyptian cubits	10 × 1.719 = 17.19
The marble gallery to the chamber	6.87	4 Egyptian cubits	4 × 1.719 = 6.876
In the middle of the gallery was a marble way its width	3.433	2 Egyptian cubits	2 × 1.719 = 3.428
The height and length of the benches of the polished marble	1.717	1 Egyptian cubit	1.719

greater irregularities observed by Mr Greaves in the best buildings of the Romans".[404] The other measurements given by Greaves can all be converted into even amounts of Royal, Memphis or simple Egyptian cubits: the main gallery was 138 English feet, or 20 Royal cubits; two other galleries were found to be 110 English feet or 16 Royal cubits; another chamber 17 English feet or 10 Memphis cubits and so on. Newton also demonstrated that measurements of the King's monument such as the height and length of the benches of the polished marble were 1.717 English feet; since a cubit is 6 palms, a palm was equal to 0.286 English feet and a digit was equal to 0.0715 of an English foot. Further measurements by Greaves, taken in English feet, proved to be equal to measurements in palms and digits based on the conversion previously mentioned and quoted by Newton. From this Newton concluded, "And it is my opinion that the Pyramid was built throughout after the measure of this (Memphis) cubit".[405]

To support the argument that the ancient buildings were built to a standard unit of measurement, Newton considered the measurement of Babylonians bricks. They were all uniform in size, according to the measurements of Samuel Purchas, their length was one foot, the width was eight inches and the thickness was six inches. So that two bricks in length, three bricks in width and four bricks in thinness formed a square cubit. Therefore, Newton concluded that the Babylonian cubit was two English feet,[406] but this is much larger that the human elbow from which the word "cubit" derived, and Newton failed to note this. Yet he still claimed that the Babylonians built in cubits. Therefore, this consistent measurement for bricks was the Babylonian cubit. He claimed that all measurements which exceeded human proportions, such as the Roman calamus, clima, scruplum, actus and many others, were deduced from the multiples of human proportions. The ancient nation rounded off their large numbers into even numbers of cubits – the cubit of man.[407]

Greaves found the modern Egyptian cubit was 1.824 English feet, exceeding that of the ancient Egyptian cubit or Memphis cubit. "The measurements of feet and cubits now exceed the proportion of the human members".[408] According to Greaves' measurements of the Egyptian monuments, the human stature was the same as it was in ancient times. The measurements have increased because of human and instrument error.

> Feet and cubits were used first (as a measurement) in every nation according to the proportion of the members of a man, from which they were taken. For the foot of a man is to the cubit or lower part of the arm of the same man as about 5 to 9.[409]

Newton confirmed this ratio of 5:9 between the foot and cubit with other ancient measurements. He considered that the Jewish measurements were determined in the same manner.

Newton claimed that Villalpando had calculated the Jewish vulgar cubit to be two and a half Roman feet,[410] but since this does not fit the "cubit of man" Newton rejected it outright. He claimed that the Jewish vulgar cubit cannot exceed the cubit of a tall man.

> The stature of the human body, according to the Talmud, contains about three cubits from the feet to the head; and if the feet be raised, and the arms lifted up, it will add one cubit more and contain four cubits. Now the ordinary stature of men, when they are bare-foot, is greater than five Roman feet, and less that six Roman feet, and may be best fixed at five feet and an half.[411]

According to the Book of Erubin in the Talmud, the area of "his place" is "three cubits for his body and one cubit to enable him to take up an object at his feet and put it down at his head".[412] Newton also moved away from the classical "Vitruvian" man. In Vitruvius the height of man is set at six Roman feet; Vitruvius claimed that the number six was perfect and this perfection was further expressed in the cubit which equalled six palms or 24 digits.[413] Newton's measurements of the stature of a man, five to six Roman feet, equalled three vulgar cubits, which was to be no less that 20 Roman unciæ and no more than 24 unciæ. With the extra cubit, the height of a man with raised arms became the sacred cubit, which he calculated to be no less that 24 Roman unciæ and no more that 28.8 unciæ.

Newton gave two examples from ancient literature, where he further defined the limits of the sacred cubit. In the first, Josephus wrote that the columns of the great court of the Jewish Temple could be embraced by three men with their arms joined. Newton claimed that the orgyia or the fathom of a man, which is the length of the outstretched arms of a man, was supposed to be the same as the height of a man but in fact is a palm wider.[414] Vitruvius stated, "For if we measure the distance from the soles of the feet to the top of the head, and then apply that measure to the outstretched arms, the breadth will be found to be the same as the height".[415] Newton further abandons this traditional image of Vitruvian man, which is confined by the circle and the square, by adding an extra palm to the length of a man's outstretched arms giving a slightly more elliptical and rectangular image to the geometry of man (see Fig. 5.2).

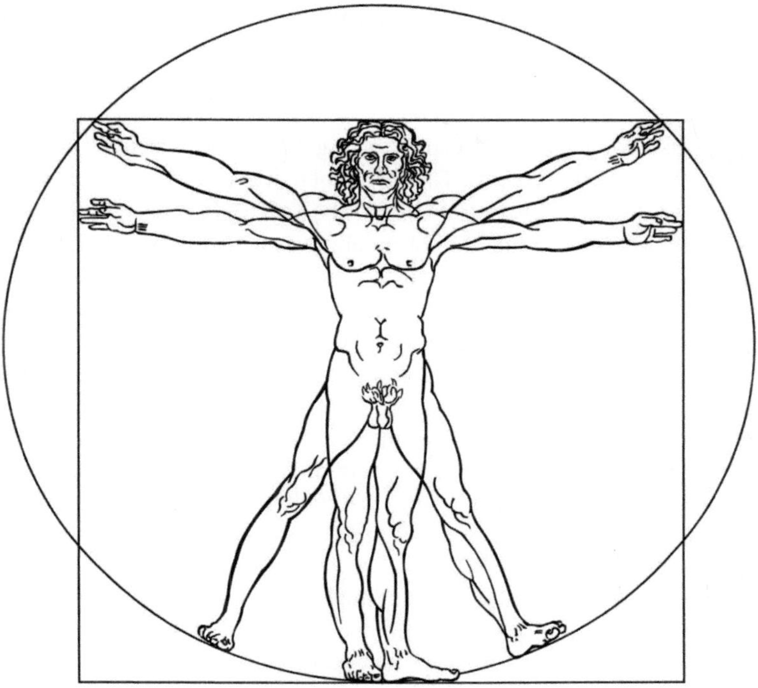

Fig. 5.2 The Newtonian man[419] (Drawn by author from Newton's description in Isaac Newton, 1737)

The circumference of the columns, according to the Talmud and Josephus is eight cubits, for Newton this is equal to three times the height of a man plus three palms i.e. greater than 15.75 Roman feet and less than 18.75 Roman feet. This further defined the sacred cubit to be greater than two Roman feet and less than two and a third Roman feet.

In Newton's second example of the use of the cubit from the ancient literature, the Sabbath day's journey, in the opinion of what Newton called the "unanimous" content of the Talmud and all the Jews, was 2,000 cubits. According to Josephus, this measurement is not so consistent, and he claimed that the Sabbath day's journey is five stades (3,000 Roman feet) and in another place six stades (3,600 Rome feet).[416] Newton, who was very familiar with the work of Josephus, used the reference from the Talmud instead and claimed that instead of "cubits" the Jews sometime substituted "paces". Walking on the Sabbath is not hurried but is of a moderated speed:

> Now man of a middling stature, in walking in this manner, go every step more than two Roman feet, and less that two and a third. And within these limits was the sacred cubit circumscribed. And within these limits was the sacred cubit circumscribed.[417]

From the height of a step from Vitruvius,[418] Newton claimed that the middling proportion referred to by the Jews was about 13.5 unciæ, and from this he calculated that a pace or sacred cubit was less than 27 unciæ and more that 24 unciæ. From the examples of the height of a man, the circumference of the columns and the Sabbath day's walk, Newton defined the limits of the sacred cubit and rejected "the erroneous opinions of other writers".

Newton argued that the vulgar cubit of the Jews was derived from the Memphis cubit when the Jews were held captive in Egypt. The vulgar cubit was used for building and daily measurements. Therefore, there was a need for a cubit for profane use and this was provided by the Babylonian cubit of two Roman feet. The proportion of the Babylonian cubit to the Memphis cubit was 6:5.0157 rounded off to 6:5. In Babson 0434, Newton confirmed that the sacred cubit was six palms[420] and the vulgar cubits was five palms.[421] In the "Dissertation", Newton continued to define his limits using Greaves measurements and a proposition by Mersennus, which defined the cubit to be 23.25 French feet. He concluded that the sacred cubit was 25.6 unciæ.

The "Dissertation" ends with a comparison of Josephus' and the Talmud's measurements, which is similar to the one in Babson 0434.[422] In Babson 0434, Newton does not state which vulgar cubit Josephus used but in the "Dissertation" he clearly states that it is the Roman cubit and that the ratio between the Roman cubit and the sacred cubit when rounded off was 2:3.

In attempting to establish the length of the Hebrew cubit Newton was endeavouring to prove the impossible. The paper is ingenious but it has problems and is fundamentally flawed. The published translation is poor and has many misprints. For example there is a misprinted number on page 411, which is corrected on page 412.[423] On page 429, the sacred cubit in the time of Moses was 25.6 in. of English feet while three pages later the sacred cubit is 25.6 unciæ of Roman feet.[424] The translation at times is confused. "And" is often printed instead of "or" as for example: "four Palms, and sixteen Digits".[425] This clearly should be "or" not "and" since four palms are equal to 16 Digits. The sacred cubit consisted of "only a single palm"[426] and this should have said "an additional palm".

What the Memphis cubit is quite unclear throughout the paper. Newton clearly stated in the beginning of the paper that a Memphis cubit was 1.732[427] English feet and worked his calculations according to this measurement. Later he claimed that the Memphis cubit was 1.719 English feet[428] and this figure was previously referred to as an Egyptian cubit, as being distinct from the Memphis cubit, and even later he claimed that "the different measurements of the cubit of Memphis, taken from the pyramid were 1.717, 1.719, and 1.732 of the English foot".[429] From Newton's defined limits of the sacred cubit and from the three Memphis measurements, he derives the sacred cubit to be 25.6 unciæ. The proportion of the sacred cubit and the vulgar cubit is 6:5 and this proportion confirmed the vulgar cubit to be 1.719 English feet. "Thus therefore, by means of these limits, those measurements agree with the sacred cubit, and consequently the measurements of the cubit of Memphis agree with the vulgar Cubit".[430]

From Greaves' measurements of the Great Pyramid, he worked out the length of the Memphis cubit. Greaves took the Memphis cubit, which he estimated to be 1.717 English feet, to consist of six palms. He took his measurements in English feet but he proved that all the other measurements of the pyramid could be measured in even amounts of palms and digits according to this estimate i.e. there would be no fractions of a digit. From this Memphis cubit, Newton derived the vulgar Jewish cubit as stated above. However, Newton claimed that the vulgar Jewish cubit consisted of five palms.[431] If this was the case, then many of Greaves' measurements would no longer equate neatly into palms and digits.

The length of a palm is very confusing in the "Dissertation". Using the measurements of the ancient writers he claimed that the Roman and Greek cubits were one and a half Roman feet and like the sacred cubit they consisted of six palms. But Newton gave the length of the Roman cubit to be 1.4504 English feet. From Greaves' measurements, Newton took the palm to be 0.286 English feet and after some calculation the sacred cubit to be 2.068 English feet. However, none of these measurements equate to each other. He clearly stated that the proportion of the sacred cubit to the vulgar Jewish cubit is 6:5 and equals six and five palms, respectively. From his final estimation of the sacred cubit, which is 2.068 English feet, this infers that the palm is equal to 0.3438 English feet but Newton retains his previous palm measurement, which he used to calculate the Memphis cubits and vulgar Jewish cubit, and to support his measurement for the sacred cubit. He retained the Talmud's description of the sacred cubit to be equal to six palms[432] and the vulgar cubit as being five palms,[433] and also Ezekiel's "The cubit is a cubit and a hand breath".[434] He also attempted to retain other ancient measurements from literary sources and Greaves' measurement of the pyramids, but the palm measurement, which is crucial to Newton's argument, is never satisfactorily resolved.

Conclusion

"Dissertation" was published 10 years after Newton's death in 1738. Although the "Dissertation" is presented as a finished work it should not be considered at all complete since it was translated directly from a working manuscript. It was originally an Appendix in "De magnitudine cubiti sacri" which is a draft on Solomon's Temple that became Babson MS 0434. In Babson MS 0434, he only repeated the section on Josephus' measurements and he does not attempt to specify the length of the cubit except to say that the sacred cubit equals six palms and the vulgar five palms. Newton does not return to the study as there are no other surviving drafts and he only used a small section of it in Babson MS 0434.[435] However, it is important to consider this paper in context with his other work on the Temple. His application of the principles of limits to these ancient measurements is ingenious and, despite its flaws, "Dissertation" is a significant manuscript. It gives an insight into Newton's vast understanding of ancient sources and measurements, his working methods and his interest in gaining a complete understanding of the Temple.

Chapter 6
An Overview of the Contents and the Source of Babson MS 0434

Overview of Babson 0434

The opening page of Babson MS 0434 is the only section that directly addresses the title of the manuscript, *Introduction to the Lexicon of the Prophets, Part Two: About the Appearance of the Jewish Temple.* Here, Newton claimed that the Hebrew texts, the legal constitutions and the Apocalypse are full of figures or symbols that cannot be understood separately, but only in conjunction with each other. He claimed "Consider the world-universe of the Israelites and the meaning of its parts and the significant of its ceremonies, which need to be explained".[436] The enacting of these ceremonies has a threefold form: the Tabernacle; the first Temple of Solomon and the Second Temple, until its destruction by the Romans in AD 70. These three sanctuaries designated different periods of religious history of prophesy.

Newton examined the measurement of the Tabernacle from Exodus 30 and 1 Kings 7. Solomon had maintained the same proportions for the area of the Tabernacle of Moses but he doubled the measurements. David had passed the plan of the Temple, which he had received from God, on to his son Solomon, but after the destruction of the Temple Ezekiel had a vision from God of this same Temple. In Ezekiel's vision, all the measurements of the Temple of Solomon had been maintained, Newton added emphatically "for which I know".[437] Later in the manuscript, he claimed that "The Temple of Solomon together with its atrium itself has not been described sufficiently in any place except in the visions of Ezekiel, whose narration is extremely difficult".[438] Throughout the text, Newton pointed out corruptions between the different texts and problems with the Book of Ezekiel, but he explained that these were because of Ezekiel's obscure language and not the vision itself.

Newton examined the changes in the Temple over time. The building of the second Temple by Zerubbabel followed the same foundations but with a great deal less grandeur. It had the same dimensions and was a pragmatic house of worship, but its architecture was mundane and it was nothing to look at.[439] Cyrus the Great ordered the building of the Temple and the internal atrium, but nothing else was added. This was the sanctuary that was maintained up to the time of Alexander the Great as reported by the pagan writer Hecataeus. The Temple was further fortified under Simeon the Just, until Herod built a more sumptuous building for the sanctuary.

T. Morrison, *Isaac Newton's Temple of Solomon and his Reconstruction of Sacred Architecture*, DOI 10.1007/978-3-0348-0046-4_6, © Springer Basel AG 2011

According to Newton, "God, predicting all these things, thus he corrected them through the prophet Ezekiel".[440]

Newton briefly examined the verses of Ezekiel 40:5–42:15 and 46:19. Verse by verse he scrutinized the dimensions of the gates and the interior and exterior atriums with their colonnades. He mapped out the floor plan of the Temple, relating it to each verse through three floor plans and he constantly checked the measurements. This examination of the dimensions is very cryptic with little or no explanation, but occasionally he pointed to inconsistencies with Kings, the Vulgate and the *Septuaginta*. After the verse-by-verse analysis of the measurements, he began examining some of the verses with inconsistencies between the *Septuaginta* and the Hebrew and Latin texts.

This is followed by a "Commentary" which is the main section of the manuscript. It begins with a description and plan of the altar, and he corrected where he perceived "the words of Ezekiel were interpreted erroneously".[441]

The description of Ezekiel was open to interpretation and was incomplete in its measurements; in Newton's "Commentary" he attempted to rationalise the measurements given by Ezekiel and the missing measurements. The architectural features were examined through the writing of ancient writers, such as Philo, Hecataeus, Josephus, Maimonides, the *Talmud* and the *Septuaginta*. He examined these writers to distinguish the additions and changes made to the design through time. He integrated the rituals into the building as another way of justifying the floor plan of the Temple. Where the descriptions of the ancient writers show that the proportions were not double that of Moses, he explained that "this happens because the Jews had shortened that atrium from the eastern part so what remained was more space for the Atrium of the Women".[442] He looked for confirmation of Ezekiel's description though these ancient writers; where Ezekiel gave overall measurements, Newton showed that the collection of buildings described by the ancient writers fitted these same dimensions (Newton, c1680).[443] A comparison of the measurements through the different stages of the development of the Temple is carefully executed and inconsistencies discarded.

Newton presented a table of measurements in which he compared both the vulgar and sacred cubits cited by Josephus to the sacred cubits cited by the experts of the *Talmud*. Thus, he demonstrated that "when the measurements of Josephus are reduced to the sacred measurements evidently they are in agreement with the ones cited by the experts in the Talmud or, better still, approach them".[444] It is only when Josephus "upon writing for the gentiles, was not careful enough with the measurement as is evident and he made mistakes".[445] Once acquainted with these mistakes of Josephus, it is possible to see how he measured the Temple.

Newton examined the colonnades: their height, the numbers of columns, their thickness, their intervals and their style. These are discerned according to the proportions of architecture.[446] From this description of the Temple, Newton claimed that it is possible to distinguish the plan of the Temple of Solomon. Since Zerubbabel had built on the foundations of the Temple of Solomon, everything that Zerubbabel and Herod added, or anything that is irregular, must be rejected. Symmetry and harmony in the design of the Temple were important factors in the layout of the

Temple plan. Newton accepted particular dimensions of Josephus on the strength of their not being harmonious with the plan.[447] He stated that "The structure is valued by such great simplicity and harmony of all its proportions".[448] The perfection of the measurements is of paramount importance to the design.

Newton then proceeded with a more extensive comment on the vision of Ezekiel. He examined verse-by-verse Ezekiel Chaps. 40, 41, 42, 43:1–7 and 46:19–24. In this examination, he again related the measurements given by Ezekiel to his floor plan with more detail. Further, he footnoted explanations of corruptions of the texts between Greek, Hebrew, Alexandrian, Arabian and Latin texts along with traditional Jewish texts. He also expanded some of the details of the text and briefly commented on other contemporary commentators such as Villalpando, Cappel and Montano. However, he also added unexplained elements such as in Chap. 42, verse 1 where he stated the number of the rooms as 15, as if this number were part of Ezekiel's text,[449] whereas Ezekiel did not mention how many rooms there were. This also contradicts his earlier claim that there were 12 rooms, which he reaffirmed in his first floor plan of the Temple.

In the final section, he stated "we complete the description of the Temple (of Solomon) comparing all the Temples between itself and supplying what Ezekiel omitted relative to the Temples of Solomon and of Herod".[450] Here, Newton used many other Biblical texts, such as 1 Kings, 1 & 2 Chronicles, Psalms, Ezra, Proverbs, as well as Josephus, Maimonides and the *Talmud* to complete the details of the first Temple – the Temple of Solomon. The manuscript finishes without a satisfactory conclusion as to how it fits into the title of the manuscript given by Newton. Babson MS 0434 is a working manuscript and, like his research on the sacred cubit, it exposes his changes in the design, and there are difference between the initial design and his final design. Despite this, the image of the Temple is explored in sufficient detail to recreate Newton's reconstruction of the Temple of Solomon.

The Sources of Babson 0434

Babson 0434 was written in four languages, the primary language being Latin, with some quotations in Greek, a few small expressions in Hebrew and one paragraph in English, which is written as marginalia in the main floor plan. He cited a wide range of Biblical texts, the Greek text *Septuaginta,* texts in Hebrew and Vulgate Latin, the Alexandrian Codex and the Arabian version. In addition he references: Flavius Josephus *Antiquitates Judaicae* (*Antiquity of the Jews*), *Bellum Judaicum* (*The Jewish Wars*) and *Contra Apionem* (*Against Apion*); Philo, *Upon the Monarchy;* Maimonides, *De Apparatu Templi* (*Apparatus Temple*) and *Tratado Sobre el culto Divino* (*Treaty upon the Divine Worship*)*;* Constantino L'Empereur, *Talmudis Babylonici Codex Middoth sive De Mensuris Templi*; Arias Montano, *De mensuris* (*Upon Them Measured*); Johannes Buxtorf, *Lexicon Talmud;* and Walton, *Bible Polyglotta.* He also mentioned Villalpando's reconstruction of the Temple of Solomon and Cappel's commentary on Villalpando in *Trisagion sive Templi delineatio triplex*

Hierosolimitani, in Brian Walton's, *Biblia Sacra Polyglotta*. Cappel and Drusius are mentioned together without any reference, and finally Vitruvius and his proportions of architecture are mentioned.

Newton's referencing of traditional Hebrew texts does bring into question his knowledge of Hebrew. Westfall claimed that Newton "learnt Hebrew in order to read Ezekiel in the original".[451] José Faur suggested that eminent Jewish scholar Isaac Abendana was Newton's Hebrew teacher[452]and that Abendana instilled into Newton his interest for Maimonides and Jewish measurements.[453] Frank E. Manuel believed that Newton could only use Hebrew with the aid of a dictionary,[454] and Mat Goldish claimed that Newton read only "some" Hebrew.[455] Although he possessed five Hebrew texts in his library,[456] he did not quote verses or passages in Hebrew from the Jewish scholar Maimonides or the *Talmud*. He only quoted small expressions in Hebrew of no more that four words, but mostly he was only emphasizing an individual word. This indicates that his understanding of Hebrew was limited and he required the aid of dictionaries and lexicons which were evident in his library.[457] On the other hand, he used the Yiddish expression "Talmudists"[458] for Talmud and the Hebraized spelling "Noach" for Noah.[459] It would appear that although Newton was familiar with Hebrew, he was never truly confident with it.

In the seventeenth century, Maimonides was the most translated and respected Jewish scholar. Christian Hebraism had become a developing interest in the sixteenth and seventeenth centuries. Hebrew learning was at the highest academic circle, and there were a large number of lexicons, grammars, dictionaries and Bibles available to these scholars.[460] Newton had a wide range of Jewish literature available to him in Latin and Greek, and the Maimonides in his library are translations into Latin by L. de Compiègne de Viel.[461] Furthermore, he used the Latin titles in his references to these books. His knowledge of the Jewish text the *Middoth*[462] appears to have come from Constantinus L'Empereur whom he referenced[463]; however, this is not in his library.

Newton made extensive use of Josephus' description of the Temple. In many commentaries and reproductions, these same passages are used to confirm various reproductions, but Newton's examination of Josephus is far more extensive than most. The measurements are considered in great detail and he compared them with the *Talmud* and equated them to Ezekiel's measurements. Josephus' measurements of the Temple are also examined in *A Dissertation upon the Sacred Cubit of the Jews,*[464] as mentioned in the previous chapter. Newton commented on the fact that Josephus and Philo had seen the Temple of Herod and had worshiped there, which gave them a better understanding of the building and the rituals while the experts of the Talmud had not seen it, which sometimes led them into error.[465]

References are made to Villalpando's reconstruction of Solomon's Temple, which was originally published in *Ezechielem Explanationes,* in 1604. From the text it appears that Newton's knowledge of Villalpando came from the criticism by Cappel in Walton's *Prolegomena of the Biblia Polyglotta,* which was to be found in Newton's library.[466] In the first part of Cappel's treatise, he included abstracts from Villalpando's reconstruction in *Ezechielem Explanationes* and he also included small-scale engravings of his design by Wenzel Hollar. *Ezechielem Explanationes* was an extremely expensive three volume set with large fold-out engraving of very high quality. Villalpando's reproduction of the Temple of Solomon became widely known through Cappel's treatise.[467] It is possible that Newton's knowledge was through the treatise by Cappel rather than through Villalpando's *Ezechielem Explanationes*.

Newton was clearly familiar with Book III chapter III "The Proportions of Intercolumniations and of Column" in *De Architectura* by Vitruvius. Yet there was no Vitruvius in his library or any other commentary on Vitruvius. Thus, it is difficult to know whether his comments on, and relating to the Vitruvian proportions,[468] are directly from Vitruvius or from one of the commentators on Vitruvius, such as Leon Battista Alberti, Sebastiano Serlio, Daniele Barbaro or any other commentary that was available in the later seventeenth century. However, he did quote Vitruvius' Book Three *De Architectura* on measurement in *A Dissertation upon the Sacred Cubit of the Jews* but, as mentioned in the previous chapter, he ignored the famous proportions that created the Vitruvian man in Book Three.

Conclusion

Newton's primary source for his reconstruction is the Book of Ezekiel. In Babson MS 0434, Newton also demonstrates his knowledge of literary sources, both ancient and contemporary, in his justification of the Biblical text. Throughout Babson MS 0434, Newton developed his plan from these sources and continually refined it. The main floor plan in Babson MS 0434 appears at the very beginning of the manuscript; however, this is a working document and in Newton's continuing refinements he made changes to this floor plan. The next chapter will consider these changes and his final, or at least the last surviving, reconstruction of the Temple.

Chapter 7
Reconstruction

The Differences in the Images of the Temple in the *Chronology* and Babson MS 0434

There are two main sets of surviving drawn plans for the Temple of Solomon by Newton[469]; the three illustrations in the *Chronology* and the six in Babson MS 0434. Of the six drawings in Babson MS 0434, three of them are incomplete and are not referred to in the text. Only four of the six drawings have sufficient detail to allow their annotations within the text to be followed. Two of these are of the gates, one of the altar and the other is the plan of the Temple precinct. The Temple precinct drawings have become synonymous with Newton's Temple. The images in the *Chronology* consist of floor plans of the Temple precinct, the Temple and the colonnades of the Temple. These images are complete floor plans, but their completeness is not backed up by the very brief and confused description given in Chapter Five of the *Chronology*. The image of the Temple precinct in Babson MS 0434 (Fig. 7.4) lacks some details, and the other images in Babson MS 0434 add nothing to complete these areas. In the two images of the Temple precinct, there are clear differences in the floor plan, but the overall appearance does seem similar. These two famous illustrations of the Temple precinct are repeatedly used to illustrate Newton's Temple.[470] However, the image in the *Chronology* is preferred over the Babson MS 0434 image in publications, since it fills in what seems to be the "grey" areas that Babson MS 0434 does not provide and it is a great deal clearer in its execution.

Although the floor plan in *Chronology* does have an initial similar appearance to Babson MS 0434, it is in fact different in many elements of the plan. In Babson MS 0434, Newton did not mention any stairs that led to the upper floors of the thirty rooms that surround the Temple. In the plan in the *Chronology*, there is a spiral staircase to the right of the main entrance and the priests would have to use the main Temple stairs to go to these chambers and to the lower floor. In Babson MS 0434[471] and the Book of Ezekiel,[472] the priests accessed the rooms on the lower floor from steps at the side of the Temple; neither Newton nor Ezekiel mentioned any stairs to the upper floors.

In the plan of the *Chronology*, these thirty chambers are a double row of fifteen rooms that surround the Temple. However, this would not fit the measurements given

T. Morrison, *Isaac Newton's Temple of Solomon and his Reconstruction of Sacred Architecture*, DOI 10.1007/978-3-0348-0046-4_7, © Springer Basel AG 2011

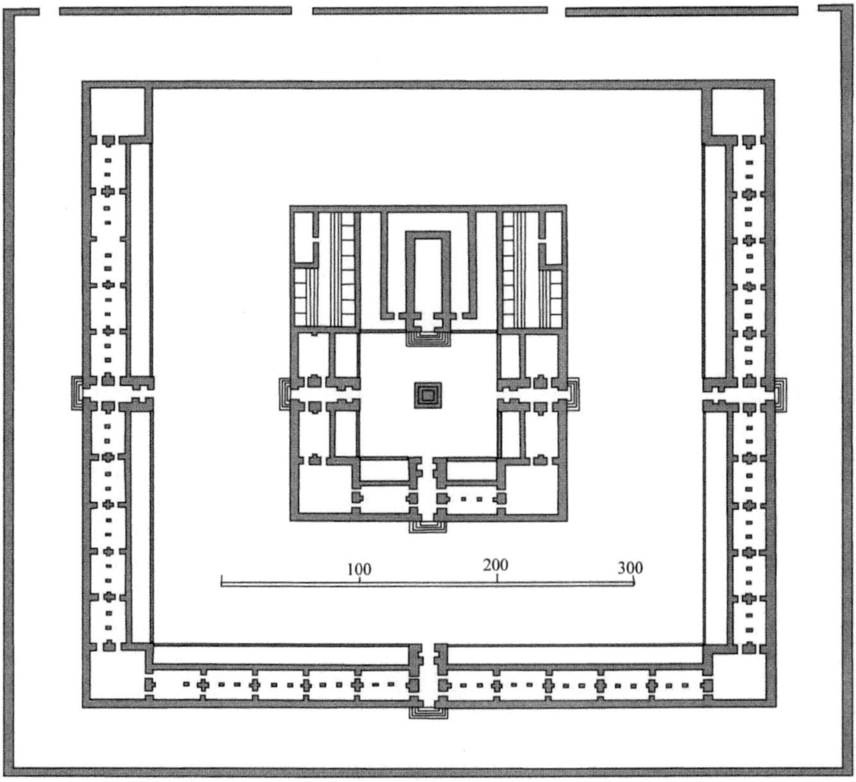

Fig. 7.1 The floor plan of the Temple precinct published in the *Chronology* in 1728[476] (Drawn by the author from Isaac Newton, *The Chronology of Ancient Kingdoms Amended* (Dublin, 1782), p. 346.)

by Newton. In Babson MS 0434, Newton clearly stated the measurement from the Temple wall to the outside of the thirty rooms as being: the Temple wall, six cubits wide; storeroom, four cubits wide; the walkway, five cubits wide; the chambers, five cubits and the outer wall, five cubits wide.[473] Newton claimed that the thirty chambers were six cubits in breadth making the circumference of the three walls of the Temple one hundred and eighty cubits,[474] the image in the *Chronology* indicates that the chambers would be at least twelve cubits in breadth. In Babson MS 0434, each chamber had its own storeroom so there were thirty chambers and thirty storerooms.

In the *Chronology*, the one hundred cubits depth of the Temple occupies the entire Separate Place, but the back wall of the Temple is also the surrounding wall of the Separate Place. This reduces the depth of the Separate Place by six cubits; a Separate Place of nine-four by one hundred cubits does not follow the proportions of Moses' measurement of the Tabernacle or the measurements of Ezekiel's text. In Babson MS 0434, this width of the surrounding wall is not counted as a part of the Separate Place.

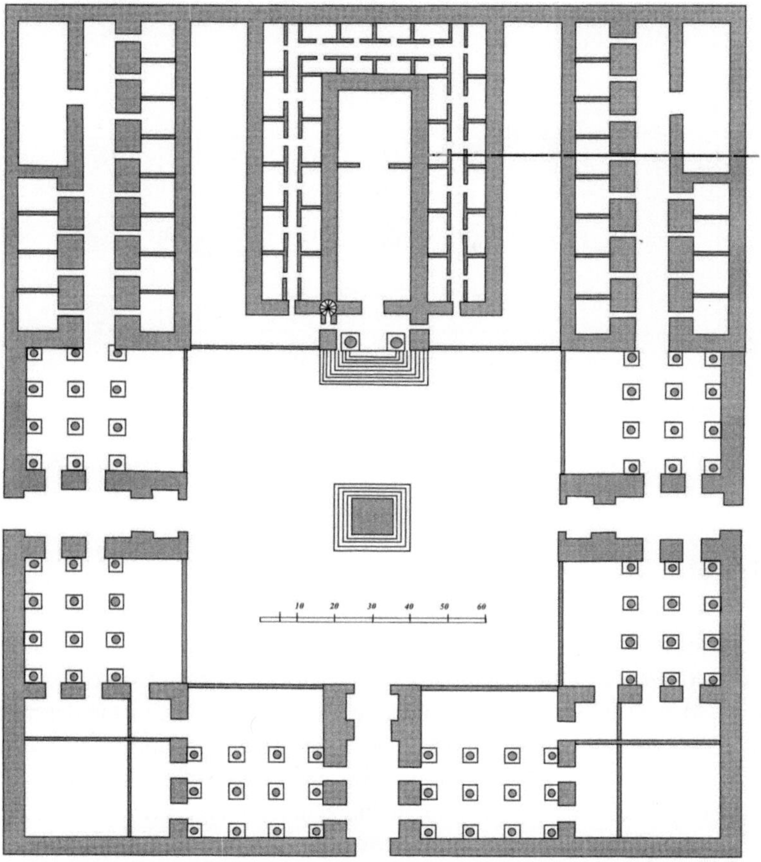

Fig. 7.2 The floor plan of the Temple published in the *Chronology* in 1728[478] (Drawn by the author from Isaac Newton, *The Chronology of Ancient Kingdoms Amended* (Dublin, 1782) p. 346.)

In the floor plan of the Temple precinct in the *Chronology*, a wall encloses the entire precinct. This wall has four gates on the western side, the Gate of Shallecheth, the Gate of Parbar and the two Gates of Assupim. This wall is not included in the floor plan of Babson MS 0434 and Newton did not attribute these gates to the Temple of Solomon but to the Second Temple.[475]

In the floor plan of the *Chronology*, both the chambers of the people in the exterior court and of the priest in the Interior Court are supported by a cloister of three rows of columns. Indeed the floor plan of this cloister is revealed, in detail, by its own illustration (see Fig. 7.3). In the floor plan of Babson MS 0434, there is a triple colonnade in front of the chambers in the exterior court, but no details are given in this image as to what is underneath the chambers. The floor plan gives no details about the colonnade in the Interior Court; however, in Newton's description of the Temple, he clearly stated that the colonnades are in front of the chambers.[477]

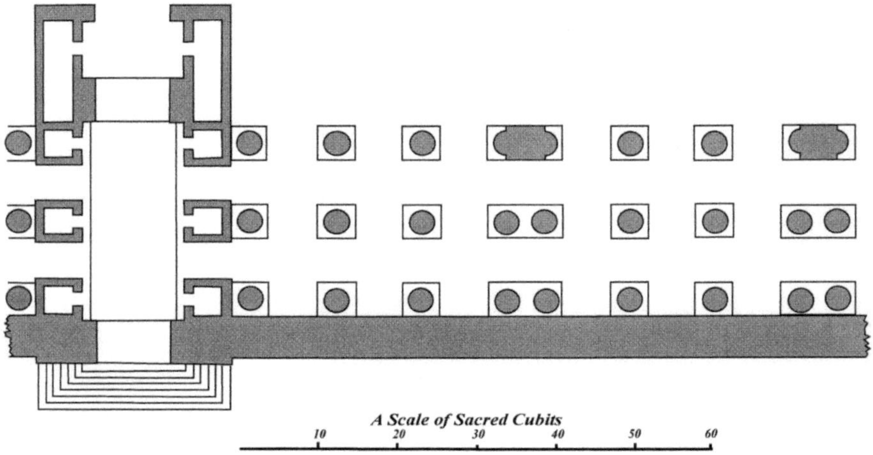

Fig. 7.3 Floor plan of the cloister under the chambers published in the *Chronology* in 1728[479] (Drawn by the author from Isaac Newton, *The Chronology of Ancient Kingdoms Amended* (Dublin, 1782) p. 346.)

The greatest visible difference in the floor plans is in the Interior Court. In the floor plan of the *Chronology*, the two eastern corners of the Interior Court are the priest's kitchens and stair cases. This is not included in the floor plan or the description of Babson MS 0434 and this area is occupied by the priest's chambers. In Babson MS 0434, the priest's kitchens are in the two western corners of the Interior Court. The stairs to the upper floors, although not included in the image, are described, but they are not in the kitchen. They are in the exterior court, they are separate and much smaller and they lead only to the upper floors.

These two two-dimensional floor plans have notable visible differences, but the differences become more extreme when the three-dimensional plan is considered.

The Differences Between the Description of the Temple in the *Chronology* and in Babson MS 0434

Chapter Five, "A description of the Temple of Solomon" in the *Chronology* is very brief and quite confused, and without the three illustrations the description would make very little sense. Even with these illustrations there are great problems. The description of the buildings is so different from Babson MS 0434 that it does raise the following questions: "is this chapter in the *Chronology* the work of Newton?"; "did he change his mind about the structure of the Temple?"; and "did he want to sanitise his work by underplaying the importance of the Temple?". The differences, the brevity, the confused description and the use of terminology that does not exist in Babson MS 0434 do strongly suggest that the plan given in the *Chronology* is not the work by Newton. On the other hand, he did attempt to sanitise some of his work at the end of his life and his editors took this even further.

In the *Chronology* it is stated that the vestibule of the Temple is one hundred and twenty cubits[480]; in Babson MS 0434 Newton claimed that the vestibule was the same height as the gates of the Great Atrium, seventy or seventy-one cubits.[481] The Temple in the *Chronology* is three floors in height, over the Holy Sanctuary the Temple was ninety cubits high and over the Holy of Holies it was sixty cubits high[482]; this strange and confused stepped structure appears to have no precedents, Biblically or otherwise. In Babson MS 0434, the height of the entire Temple, with the exception of the vestibule, is one hundred and twenty cubits.[483]

There are expressions used in the *Chronology* which are not used in Babson MS 0434 or any other unpublished manuscript, such as the "people's Court" for the exterior court, and the "cloister" for the columns under the chambers. The gate on the eastern side of the exterior court is the gate where Ezekiel entered the Temple precinct and in Babson MS 0434 Newton referred to it simply as the eastern gate, but in the *Chronology* it is referred to by the more secular name the King's Gate.[484] This expression is not used in Babson MS 0434.

In the *Chronology*, he stated that "the cubit was about 21½, or almost 22 inches of the English foot, being the sacred cubit of the Jews, which was a hand-breadth, or the sixth part of its length bigger than the common cubit".[485] This measurement contradicts Newton's earlier work on the sacred cubit. He estimated the sacred cubit to be 2.068 English feet or 24.816 inches.[486] The measurement of 21.5 or almost 22 inches given in the *Chronology* was established by Bishop Cumberland (1631–1718) and was quoted as being the measurement in numerous lexicons and dictionaries of Freemasonry.[487] A measurement of 21.5, or almost 22 inches of the English foot, would make the vulgar Hebrew cubit 1.527 English feet or 18.33 inches. This measurement would have no connection to any of the Egyptian, Memphis or Babylonian cubits which were the starting point of Newton's original estimation.

However, all these changes Newton did make in a manuscript held at Cambridge University Library, Additional MS 3988, which was written towards the very end of his life. He also added another floor plan; although it has even less details than his other floor plans it is a confused mixture of the features from the first and second temples. This is possibly where the artist got the design for the floor plan of the precinct for the *Chronology*. But the artist used a great deal of imagination by adding details that are not mentioned by Newton. Thus, the floor plans in the *Chronology* could not be considered Newton's design. Although the text in Additional MS 3988 is the same as the *Chronology*, it shows no notes or corrections by Conduitt, as do the other manuscripts, in the course of editing. So it is possible that there are other unknown or lost manuscripts on the Temple with further diagrams. Questions of why Newton was playing down the role that he had given the Temple all his life cannot really be answered satisfactorily. But with the controversy of the "Short Chronology" still raging at the end of his life, perhaps he considered it to be one way of sanctifying his work without too many questions.

The Description(s) of the Temple in Babson MS 0434

In Babson MS 0434 Newton continued to refine his plan. The text by Ezekiel is not always consistently applied. In Newton's prophetic work, he used the four-faced cherubim as an analogy of the four evangelists described in Ezekiel 1:10, but in his reconstruction he used the two-faced cherubim as described in Ezekiel 41:18. Also the text of Ezekiel is not always clear and at times his measurements did not fit the overall plan he presented. Newton attempted to make sense of the plan by Ezekiel. In certain unclear passages of the text by Ezekiel, Newton had a couple of attempts to resolve the uncertainties. However, the changes to the plan throughout the manuscript do not constitute any significant change in the ideas behind the plan; they are only refinements of the plan.

In the floor plan, there is a room marked Φ (see Fig. 7.4 and Fig. 2b, "Translation of Babson MS 0434") in the Interior Court to the west of the northern gate; another room is drawn in the same position opposite to the court. In his first verse-by-verse explanation of the text by Ezekiel, he claimed that

> Next to this northern gate was the atrium **RμmN**, and in the room **Φ** destined for the Priests that took charge of the custody of the Temple, and a similar room there were next to the southern gate for the Priests that had the custody of the altar. These rooms faced towards the most interior atrium.[488]

The position on the plan indicates that this room is positioned on the ground floor and separate from the other rooms. The text is repeated again with more detail and with footnotes included in the second verse-byverse explanation of the text by Ezekiel. However, this time the description placed the rooms of the priest that guard the Temple with the rooms of the priest of the Curias[489] i.e. the rooms above the colonnade. In the final floor plan of Babson MS 0434, there is no room on the pavement of the Interior Court and the triple colonnade of the northern, eastern and southern sides of the Interior Court is only interrupted by the three gates.[490]

In his verse-by-verse explanation, Newton had problems with Ezekiel 41:10; "And between the chambers was the wideness of twenty cubits round about the house on every side". However, there is only fifteen cubits between the side of the Temple and the side of the chambers of the priest in the Separate Place according to Ezekiel's own description. Newton made two attempts to reconcile Ezekiel's words with a consistent plan; both were unsuccessful. First, he claimed that the twenty cubits refers to the thickness of **wh,** which is the building surrounding the Temple, i.e. the storeroom + walkway + rooms + wall.[491] However, this actually measures nineteen cubits according to Newton's plan and a further problem is that he discounted the fifteen cubits pavement that surrounds the Temple so that there are thirty-four cubits between the side of the Temple and the side of the chambers of the priests in the Separate Place. In the second attempt, he claimed that the twenty cubits is not the remaining space between the chambers, as some have imagined, but it is "here the width was that of the Separate Place that separates the side

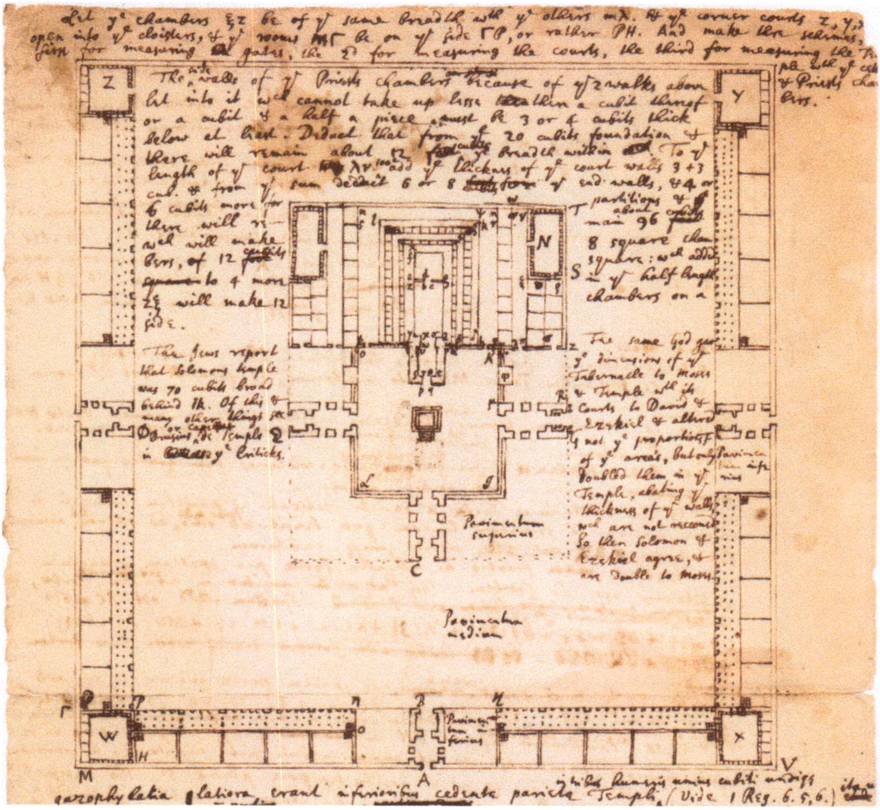

Fig. 7.4 The Temple Precinct as drawn by Newton in Babson MS 0434[500] (From Isaac Newton, *Introduction to the Lexicon of the Prophets, Part Two: About the Appearance of the Jewish Temple* (Babson Ms 0434) (unpublished manuscript, Babson College) fol. 9r, with kind permission of The Huntington Library.)

building and the chambers of the Priests where the sacrifices are consumed". [492] He calculated it thus

> if all the previously said widths of seventy cubits (the width of the Temple plus the walls) is subtracted, the width of the Temple of twenty cubits and the width of their wall of five cubits of the side and five cubits of the other, to the first reduction, they will remain forty cubits, twenty on one side and twenty on the other as the width of perimeter of the adjacent building to the Temple. Or thus the width of the side chamber is five cubits, as above it. That of the remaining space is of five cubits.[493]

Newton justified Ezekiel's measurements, ensuring that his floor plan of the Temple complied with the Prophets words.

Ezekiel 42 described the priests' chambers towards the north-west and the south-west corners of the Separate place. Although Ezekiel measured the chamber, he did not state how many chambers there were. In Chapter Two of *Observations upon the*

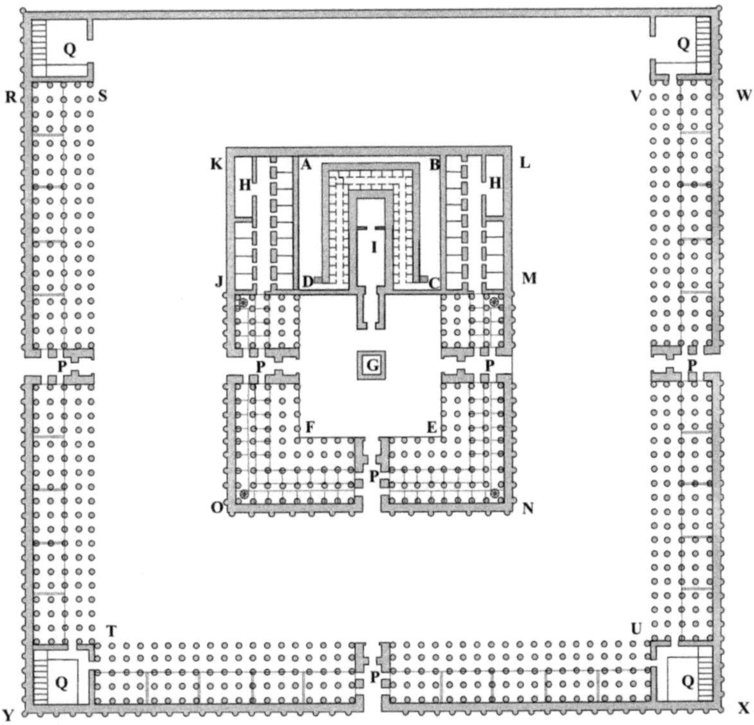

Fig. 7.5 The final floor plan of Babson MS 0434[501] (Drawn by the author from Isaac Newton's description in Babson Ms 0434.)
ABCD: The Separate Place that contains the Temple
DCEF: The Court of the Priests or the Inner Court
G: The Altar
H: The kitchen of the priest and where the sacrifices were prepared
I: The Temple
KADJ and BLMC: The chambers of the higher ranking priests
JDFECMNO: The chambers of the priests
P: The gates
Q: The kitchens of the people and the stairs to the upper chambers
RSTUVWXY: The chambers of the people of the Outer court

Apocalypse of St John, Newton placed John the Divine's vision of the Apocalypse in the Temple.

> John saw the door of the Temple opened; (and John saw the throne of God). And round about the throne were four and twenty seats; answering to the chambers of the four and twenty princes of the Priests, twelve on the south side, and twelve on the north side of the Priests Court.[494]

In Babson MS 0434 Newton detailed the use of the twenty-four chambers.[495] In the floor plan, these twenty-four rooms are represented along with a justification of

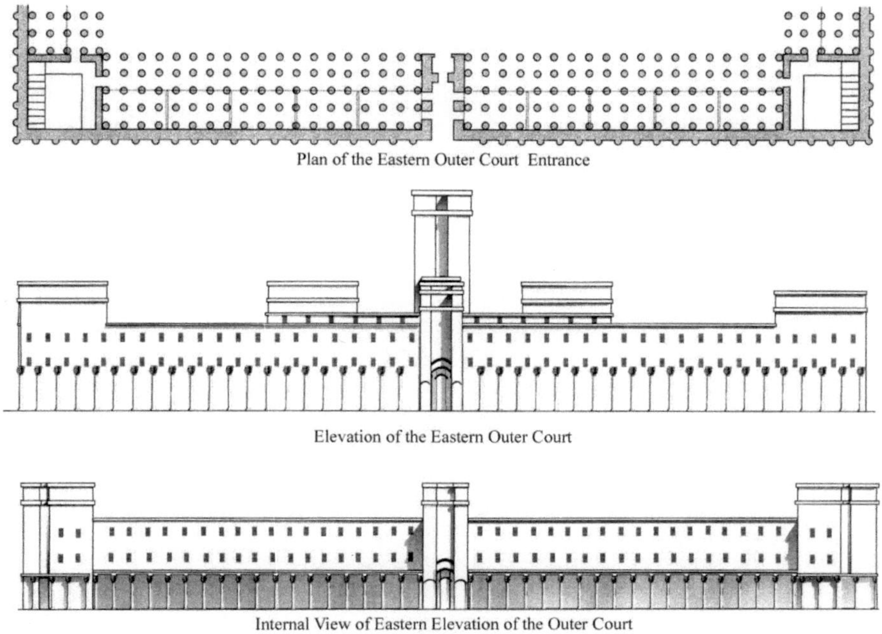

Fig. 7.6 Elevations of the Outer Eastern Court Entrance[502] (Drawn by the author from Isaac Newton's description in Babson Ms 0434.)

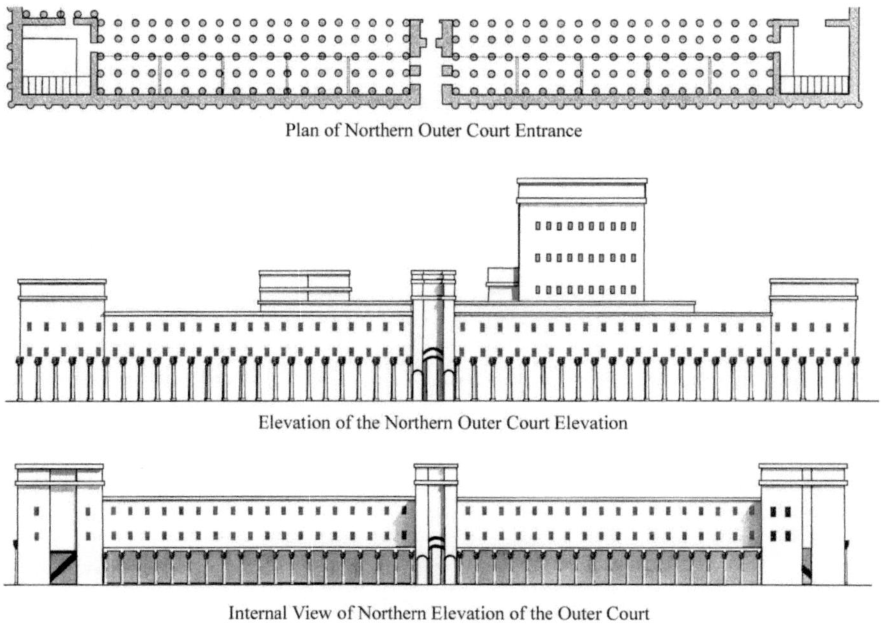

Fig. 7.7 Elevations of the Outer Northern Court Entrance[503] (Drawn by the author from Isaac Newton's description in Babson Ms 0434.)

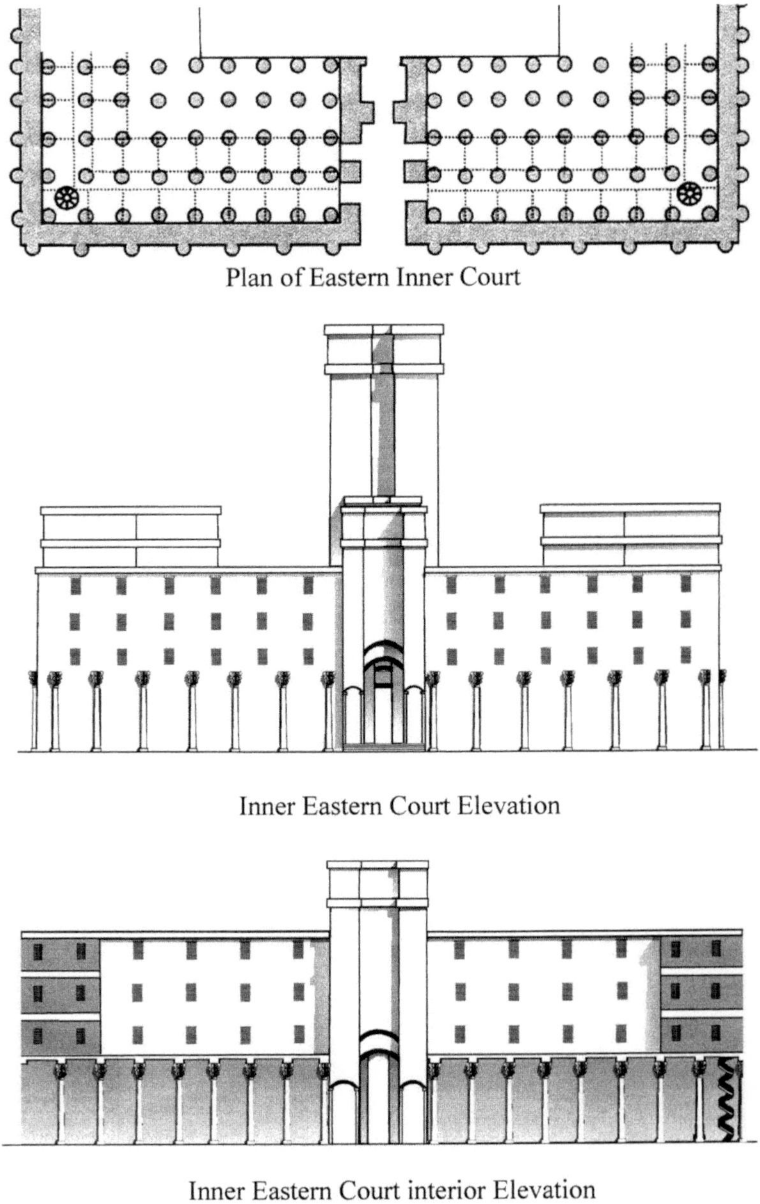

Fig. 7.8 Elevations of the Inner Eastern Court Entrance[504] (Drawn by the author from Isaac Newton's description in Babson Ms 0434.)

the measurements of these rooms.[496] These twenty-four chambers are deeply enmeshed in his ideals of the apocalypse and the ritual surrounding the apocalypse. Yet in his second verse-by-verse explanation, in his description for the floor plan there are fifteen chambers. However, this is clearly an error since at the end of Babson

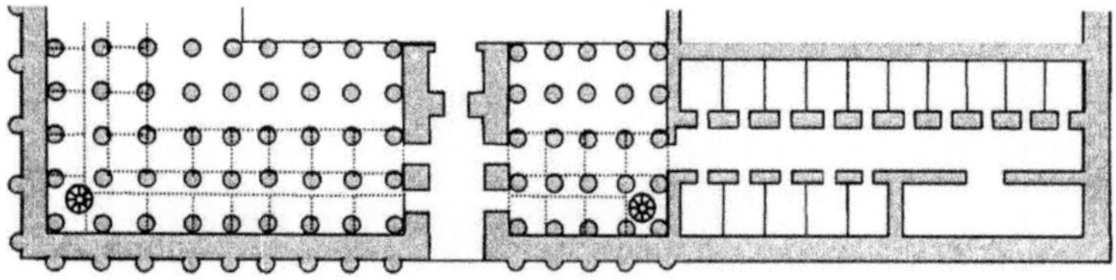

Plan of the Northern Inner Court

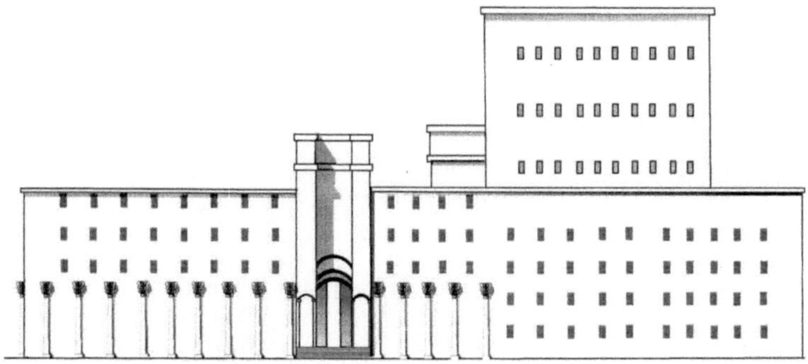

Elevation of the Northern Inner Court

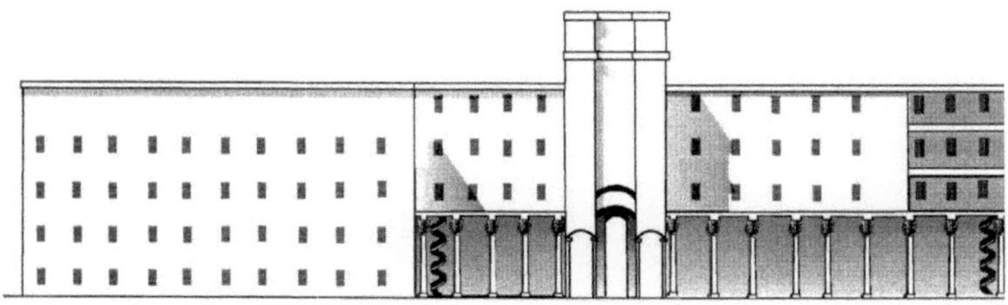

Elevation of the interior of the Northern Inner Court

Fig. 7.9 Elevations of the Inner Northern Court Entrance[505] (Drawn by the author from Isaac Newton's description in Babson Ms 0434.)

MS 0434 the rooms are being assigned to the hierarchy of priests and there are twelve rooms on each level.[497]

The Reconstruction

The reconstruction presented in this book is based on the final description of the Temple presented in Babson MS 0434. Newton omitted some details, for example, the stair to the upper chamber that surrounds the Temple, and there are also no stairs to the upper floor of the Temple itself. Another problem is that it is not always clear

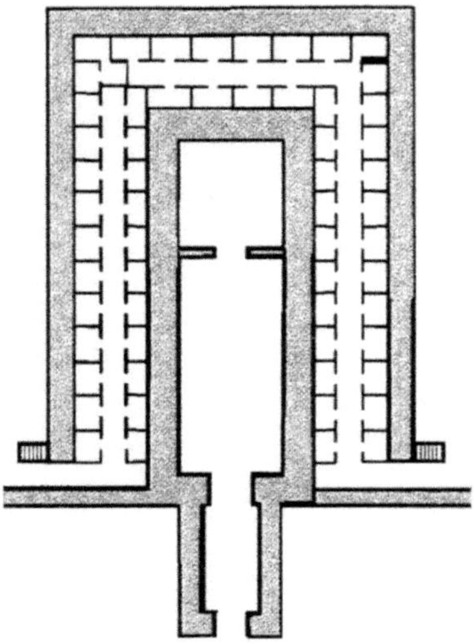

Plan of the Temple

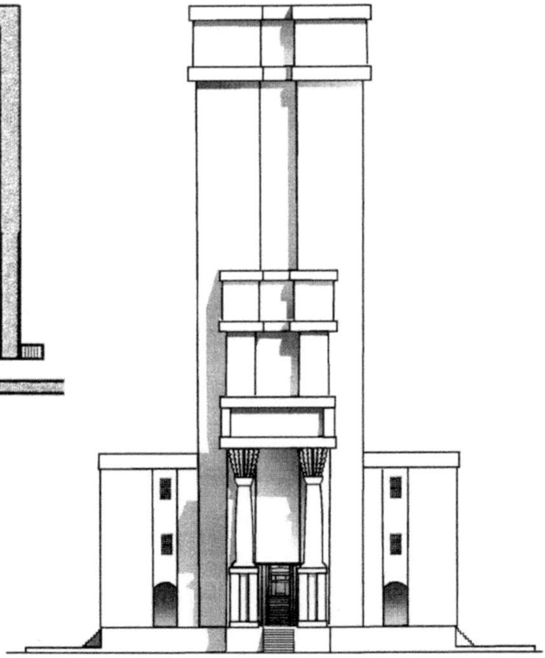

Eastern Elevation of the Temple

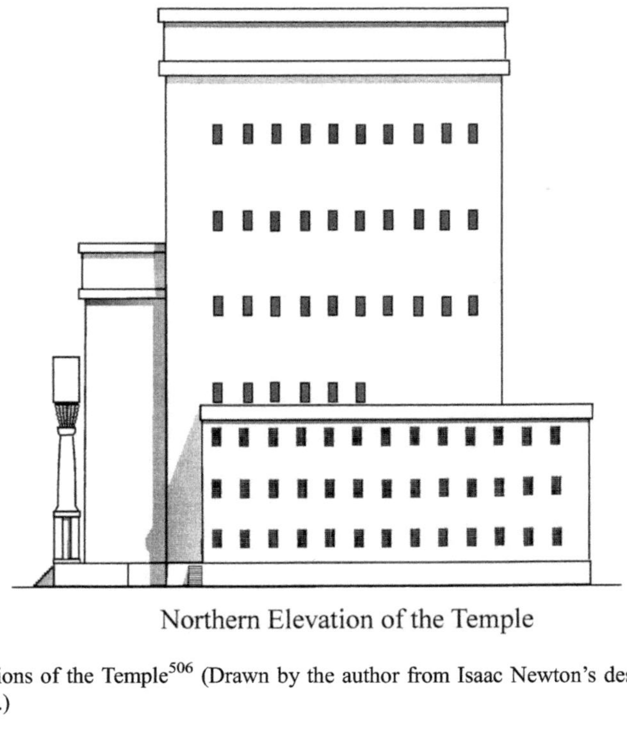

Northern Elevation of the Temple

Fig. 7.10 Elevations of the Temple[506] (Drawn by the author from Isaac Newton's description in Babson Ms 0434.)

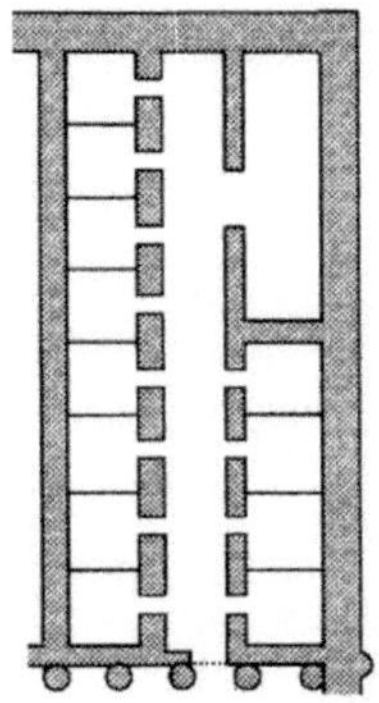

Plan of the Building for the Priests in the Separate Place

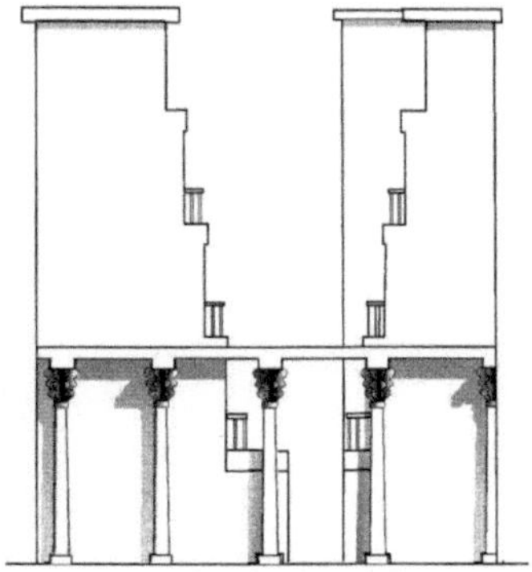

Eastern Section of the Building of the Priests

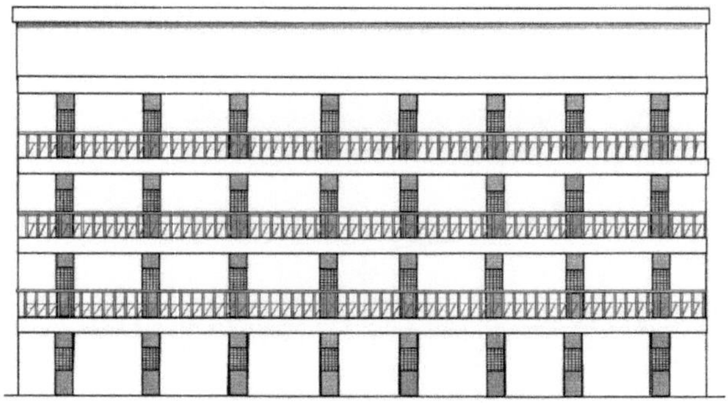

Northern Section of the Building of the Priests

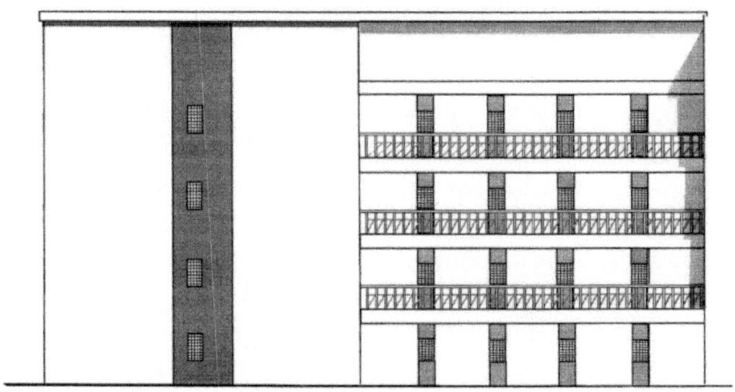

Southern Section of the Building of the Priests

Fig. 7.11 Sections of the buildings of the Priests[507] (Drawn by the author from Isaac Newton's description in Babson Ms 0434.)

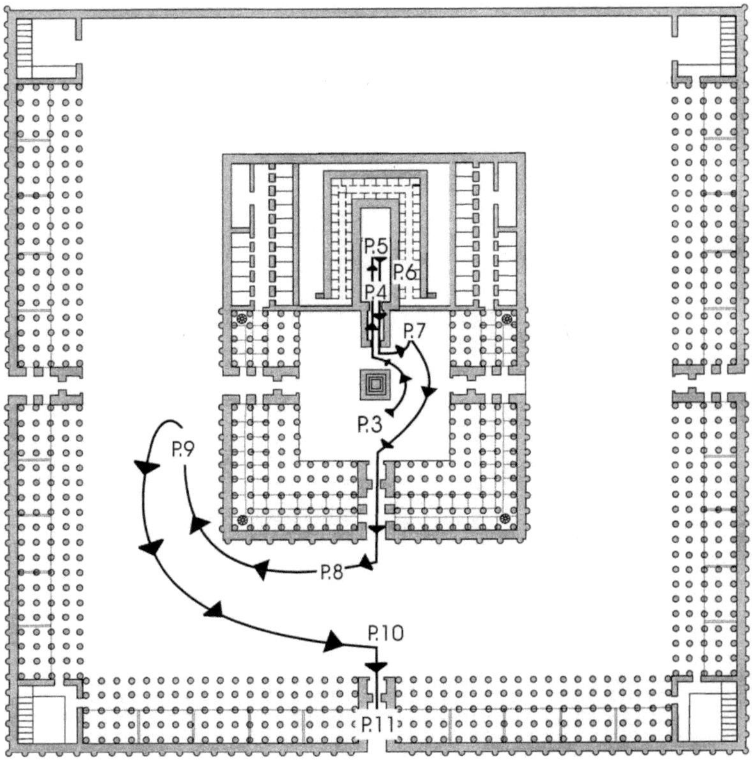

Fig. 7.12 The colour plates on pages 95–100 represent a walk through the Temple precinct[508] (Drawn by the author to describe the path of the coloured images)

whether the thickness of the surrounding walls is included in some of his measurements; with the gates the exterior wall is included in the fifty cubits but the wall behind the Temple is excluded from the one hundred cubits of the Separate Place. Despite these problems, the description is sufficient to construct the form of the Temple, which was Newton's intention.

One confusing aspect of Newton's description concerns the columns and the colonnades. Some of this confusion stems from the diagrams of the *Chronology,* which specify a cloister of three rows of columns. It also stems from the fact that Newton referred to the chambers built on rows of columns and a colonnade built in front of the chambers. In fact, they are two different construction elements that sit side by side.

Newton quoted Ezra 6:4 which states that the Temple was built "with three rows of great stones, and a row of new timber". Interpreters of the verse have translated this to mean that the rows of stones represent the floors of the building which are separated by means of cedar timbers.[498] Newton interpreted this as

> those three rows were of cylindrical stones of form, that is to say, columns. There were two rows of columns in the colonnade under the rooms; the third was in the external facade of the

> exterior wall of the rooms these corresponded with the columns of the colonnade of the Great Atrium. The series of beams of wood was found in the panelled ceiling of the colonnade,[499]

There are three rows of columns of the colonnade plus two rows that support the chambers, making five rows altogether in the court and another row of columns on the exterior wall.

Conclusion

The detailed illustrations of the Temple in the *Chronology* are not supported by the brief and confused description of its structure. The text of the *Chronology* is indeed Newton's, but is very confused and parts of it make little sense. Although the floor plan does have similarities in its overall appearance, it is very different from the plan presented in Babson MS 0434.

Newton strove to illustrate and understand what he perceived to be the God-given plan that Solomon implemented and that Ezekiel described for posterity. He also attempted to justify the text of Ezekiel and to prove that the plan and the measurements that Ezekiel gave were correct. Babson MS 0434 revealed his continual refinements of that plan. The manuscript has been dated to the mid-1680s and may not be the last development in his refinements. However, although he remained interested in the Temple throughout his life, Babson MS 0434 was the only surviving plan where it was possible to reconstruct the Temple from the outlined description.

This reconstruction of Newton's Temple follows his description as closely as possible and where Newton only gave a minimal description, i.e. measurement and the number of floors, the building is kept as the basic structure and no extra elements are added.

The reconstruction first shows the overall floor plan that was developed from the final description given in Babson MS 0434. There are differences from his first drawing, but they are minor. The main difference is that the detail in his description has been added. Second, the elevations and sections of the entire Temple are illustrated. Finally, a series of coloured plates are organised as a walk through the Temple precinct, beginning from a bird's-eye perspective of the Temple, moving into the Temple, then to its Sacred centre, around the precinct grounds and finally out through the eastern gate.

Transcription of the English Annotation in Fig. 7.4

Let ye chambers **ζz** be of ye same breadth with ye other **mλ** and ye corner courts **Z,Y,X,W** open into ye cloisters, and ye rooms **MΓ** be on ye side **ΓP**, or rather **PH** and make there schemas, the first for measuring gates the second for measuring the courts, the third for measuring the Temple with ye cells and Priests chambers.

The side walls of ye priests chambers because of ye two walks above let into it which cannot take up less there a cubit thereof of a cubit and a half a piece must be 3 or 4 cubits thick below at least. Deduced that from ye 20 cubits foundation and there will remain about 12 cubits breath within to ye length of ye court **λv** 100 (cubits) add ye thickness of ye court walls 3 + 3 cubits and from ye sum deduct 6 or 8 cubits from ye end walls and 4 or 6 cubits more for partitions and there will remain about 96 cubits which will make 8 square chambers of twelve cubits square: which added to 4 more in ye half length **zζ** will make 12 chambers on a side.

The Jews report that Solomon's Temple was 70 cubits broad behind **lk**. Of this and many other things see Drusius & Cappel *de Temple* contrary to the critics.

The same God gave ye dimensions of ye Tabernacle to Moses and Temple with its courts to David and Ezekiel and altered not y^{e} proportions of y^{e} areas, but only doubled them in y^{e} Temple, abating y^{e} thickness of y^{e} walls wch were not recounted. So then Solomon and Ezekiel agreed, and were double to Moses.

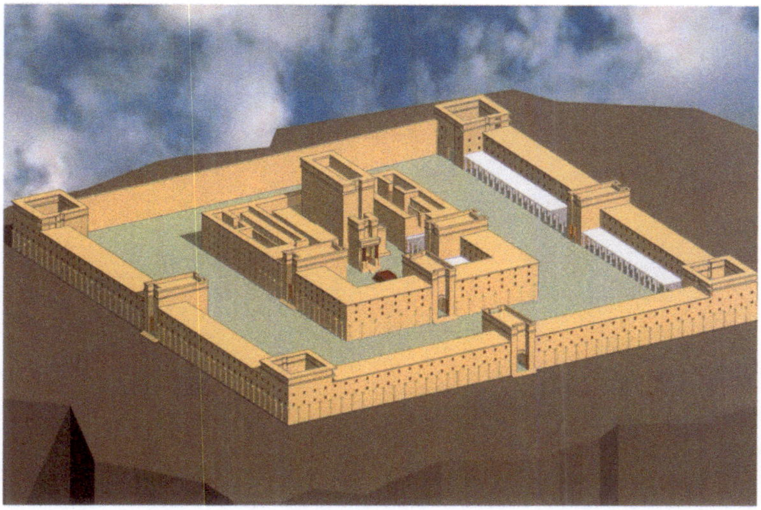

Plate 1 Birds-eye perspective of the Temple Mount

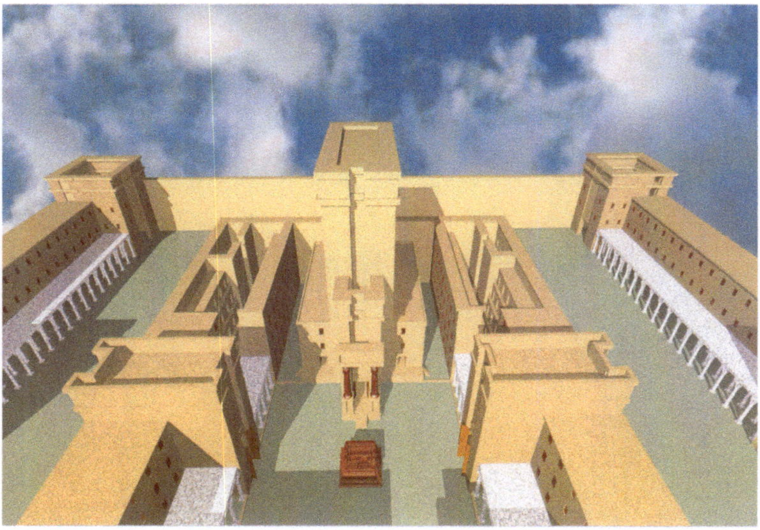

Plate 2 Birds-eye view of the Temple and its surrounding courts

Plate 3 The beginning of the walk through the Temple precinct and it begins in front of the Altar and the Temple

Plate 4 The first chamber of the Temple, decorated with cherubim and palm trees and the back of this chamber is the stairs to the Holy of Holies

Plate 5 The most sacred part of the Temple, the Holy of Holies, which contains the Ark of the Covenant. This room is only entered once a year by the high priest

Plate 6 A section of the two chambers of the Temple

Plate 7 The Temple looking at the inner court

Plate 8 The outer court turning right

Plate 9 The corner in the outer court

Plate 10 Turn back to the eastern gate

Plate 11 A section of the eastern gate. A high priest is entering at the point at which Ezekiel first entered the Temple precinct

Chapter 8
Conclusion

From a twenty-first century perspective, chronology, the interpretation of Biblical prophecy, alchemy and the Temple of Solomon appear to be strange topics for one of the great scientists that history has known to spend his time studying – let alone a great deal of time. Yet these topics were widely studied in the seventeenth and eighteenth centuries by many of the intellectuals of the day. Newton studied these topics from his early days in Cambridge and clear influences can be seen from, Sanderson, Mede, More, Villalpando, Vossius, Maimonides and many others. Although Newton's work does reveal these influences, he did not follow any one in particular. He studied the sources closely and questioned them, and this can particularly be seen in the critical comments that Newton wrote in the margins of More's books and his comments on Villalpando. Keynes considered that Newton was a "Judaic monotheist of the school of Maimonides"[509]; however, although Newton was influenced by Maimonides those influences are not as clear as Keynes suggested nor does that influence stretch very far. This is demonstrated by Newton and Maimonides' extremely different floor plans of the Temple, among other things. Above all, and throughout his life Newton searched for the truth, and for Newton that truth began and ended with God.

From at least the early 1680s, the Temple of Solomon was a part of his studies. His study of the Temple of Solomon was not just a single manuscript; the Temple was an important element of his studies of chronology and Biblical prophecy; it represented the microcosm of the macrocosm.

The Temple of Solomon as the microcosm of the macrocosm was not unique to Newton. Newton built on a long tradition that stretched from antiquity to his time. Villalpando's *Ezechielem Explanationes* expressed this tradition; in it the floor plan of the Temple was represented as a map of the universe. For Villalpando the Temple represented the perfect vision of God's Creation – a geocentric universe. The measurements and proportions of the Temple were according to the harmonies of music. This was the perfect vision of the macrocosm. Although Newton built on this long tradition, for Newton the floor plan of the Temple held the ancient knowledge of the heliocentric universe. It was more than the vision of the macrocosm; this ancient knowledge held the secrets of the workings of the universe. Understanding the mechanics of the universe through the Temple was to come closer to the understanding of God.

T. Morrison, *Isaac Newton's Temple of Solomon and his Reconstruction of Sacred Architecture*, DOI 10.1007/978-3-0348-0046-4_8,

The Prytanæum mapped this frame of the world, nature and the entire universe that was God's Temple. Thus, knowledge of God, and consequently nature, could be gained through the understanding of the plan of the Prytanæum. The Temple of Solomon was one of these Prytanæums. The prophet Ezekiel encoded the plan of Solomon's Temple into his text and with the corruption of the original religion that resulted in the demise of these Temples, his encoded text remained. Ezekiel's description of the Temple held the ancient truth. Within the architecture of the Temple was encoded the heliocentric planetary motions, the esoteric knowledge of the universe. Ezekiel wrote in the encoded language of the prophets. Newton attempted to decode this language by understanding its measurements and architecture.

Babson MS 0434 was written in the 1680s at a time of frenetic activity for Newton and at the height of his powers. At the same time, Newton was writing the *Principia,* he was also working on chronology, alchemy, interpreting the prophets and the Temple of Solomon. However, his writings cannot easily be classified into these topics because they merge into each other and the divide is often very blurred. Figure 8.1 divides Newton works into three main topics: alchemy, natural philosophy and theology. The classification of the papers is taken from The Newton Project,[510] but even in these seemingly distinct categories there are difficulties. For instance is *De Gravitatione et aequipondio Fluidorum*[511] natural history or theology; is *Of Natures obvious Laws & Processes in Vegetation*[512] alchemy or natural philosophy; is "*The Book of Nicholas Flamel conteining The explication of the Hieroglyphical Figures wch he caused to be put in the Church of the SS. Innocents at Paris*"[513] alchemy or theology, and so on. Figure 8.1 does represent a simplified division, but in reality that division is blurred. Nevertheless, it is clear from Fig. 8.1 that the 1680s were

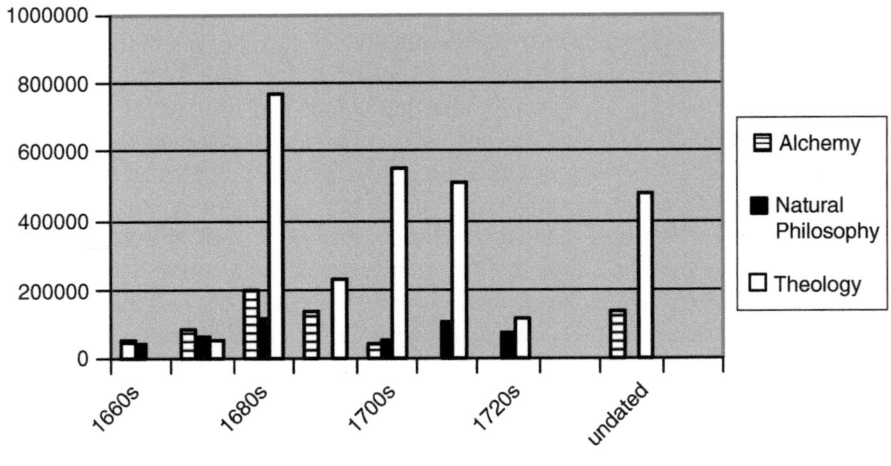

Fig. 8.1 The columns represent the number of words written by Newton from the 1660s when he began his studies at the University of Cambridge. Not represented in this figure are Newton's Mint writings; there are 916 surviving Mint documents,[515] which date from the mid-1690s to the last years of his life. These are receipts, instructions, lists of obligations and general correspondence and as such do not constitute a body of research therefore they have been excluded

Newton's most productive years in all three categories. It is also clear, contrary to popular opinion, that Newton's theological work, including chronology and prophetic interpretation was not the work of an elderly and senile Newton.[514] The bulk of Newton's theological writings was executed in the 1680s and much of the later writings was a reworking of these earlier texts.

Another common perception about Newton is that theology was a diversion from his "real" work – natural philosophy. This is clearly not the case. However, it would be wrong to state the converse that "theology was Newton's 'real' work and that natural philosophy was the diversion". For Newton, they were the two sides, the esoteric and exoteric, of the one thing, God's real Temple – the universe.

Newton's writings on prophetic interpretation, the Temple and chronology become more conservative towards the end of his life. He began to consider chapter structures that suggested that he may have been sanitising these works for publication. If he was, then his posthumous editors clearly did not think that he sanitised enough and they made further cuts to his work; the *Chronology* and *Observations* are a fraction of the work on chronology and prophecy that Newton left.

In these published books, the Temple is incidental. In the *Chronology,* although there are many references to the importance of the Kingdom of the Israelites and the reign of Solomon as the measure of time, the Temple has a small chapter of less that 3,000 words. The description of the Temple is confused and this small chapter is placed after the kingdom that destroyed it. While in *Observations* the Temple is given more importance as the "scene of the vision" of the Prophet John the Divine and the rituals of the apocalypse performed in the Temple, the significance of the Temple as the frame of the world is not explained.

In the unpublished manuscripts, the Temple serves a dominating role. It has a role at the beginning of time and religion. It contains and preserves the ancient knowledge of natural philosophy. It reveals the God-given plan of his Temple – the universe. To understand the plan of the Temple is to understand the plan of the universe. The prophets Ezekiel, Daniel and John recognised the importance of the Temple and encoded it into their texts. Its role extends to the end of time the rituals of the apocalypse that were not only performed in the Temple; the architecture was an important and essential part of these rituals.

Newton was a highly complex man and in writing about one element of his thought there is always a danger of oversimplification. However, the Temple was a part of that complexity, it was not just a study for its own sake as has been suggested[516]; it was part of Newton's philosophy and as such deserves to be considered as central to an understanding of Newton the person and Newton the scholar.

Translation of Babson MS 0434

Introduction to the Lexicon of the Prophets, Part Two: About the Appearance of the Jewish Temple

Generally, it is agreed that the future is foretold in the legal constitutions (the Hebrew texts) and the Apostle Paul bears witness to this in his writings, Colossians 2 and Hebrews 9.23. Thereupon these constitutions are more suitable than the natural world from which the prophets might choose the figures, and the Apocalypse is full of these sort of figures and these constitutions and those of the Apocalypse are thus like twins, since they prophesy from the same two matters, they explain themselves mutually, they may not be understood apart. It is in fact a legal and sealed up book (The Torah) at hand for Him who is seated on the throne and its seals are undone in the Apocalypse. Consider the world-universe of the Israelites and the meaning of its parts and the significance of its ceremonies, which need to be explained. And before all things concerning the law the sanctuaries, where these laws are enacted and they have a three-fold form: From the Tabernacle until the time of Solomon; the first Temple until the time of captivity by the Babylonians and the time of the Second Temple until the captivity under the Romans. It requires us to become acquainted with the form of these three if we wish to follow the correct meaning.

The sacred place of the Tabernacle was ten cubits in width and as many in length, the holy place was ten cubits in width and twenty in length, the altar of gold was a cubit in length and in width, the great altar was five cubits in length and as many wide, and the atrium of the Tabernacle was fifty cubits wide and a hundred cubits in length. Imagine that this atrium was divided by a line transversal in two quadrants, and that the Tabernacle was in the western quadrant and the altar surely in the centre of the eastern quadrant. To the eastern quadrant it will be called, in memory of the destruction, the Atrium of the Priests and to the other quadrant in which there was the Tabernacle itself was called the Separate Place, because this was what it was called when the Temple existed. In the centre of the Sacred Place there was the Ark of the Covenant. In the centre of the entire Tabernacle was the Altar of Gold, a partial curtain was in a distance between the Altar and the Ark of the Covenant. In the middle of the Holy Place, towards the southern side was placed the seven-fold candelabra and the golden table in front, towards the northern side. Finally, the basin or the Sea of

T. Morrison, *Isaac Newton's Temple of Solomon and his Reconstruction of Sacred Architecture*, DOI 10.1007/978-3-0348-0046-4, © Springer Basel AG 2011

Bronze was in the Atrium of the Priests between the Tabernacle and the Altar towards the south (Exodus 30: 18; 1 Kings 7: 39).

Solomon maintained the proportions of the areas of Moses in the construction of the Temple, but he doubled the measurements. And this same construction that God had revealed to Solomon through David (1 Chronicles 28: 19), after the destruction of the Temple, it was shown to Ezekiel by this same God that all these measurements had been maintained, for which I know. And thus the Sacred Place of the first Temple was twenty cubits in length and as many wide (1 Kings 6: 20; Ezekiel 41: 4). The Holy Place was twenty cubits in length and forty cubits wide (1 Kings 6, 2.17: Ezekiel 41: 2), the Altar of Gold was two cubits in length and as many wide (Ezekiel 41: 22).[517] Under Solomon, the Oracle of Moses was refurbished with new materials, as can be read in (1 Kings 6: 20).[518] The great altar, by the upper part in the circuit of the Place of the Fire, was ten cubits in length and in width, but by the lower part was double in length and in width (2 Chronicles 4: 1; Ezekiel 43). The Atrium of the Priests was a hundred cubits in length and a hundred wide (Ezekiel 40: 47), the Separate Place likewise was a hundred cubits in length and in width (Ezekiel 41: 13, 14, 15). For that reason both atria were joined, for the Atrium of the Temple that corresponded to the Atrium of the Tabernacle was a hundred cubits in width and two hundred in length. The vestibule in front of the Temple built by Solomon was twenty cubits in length and approximately ten, or more exactly eleven cubits in width (1 Kings 6: 3; Ezekiel 40: 49).

The atrium surrounded the vestibule with a splendid building and subsequently to a greater distance, with another atrium still more splendid, intervening from every direction a space of the exterior atrium of almost a hundred cubits in width. Thus, there were two atriums built; one more internal of the Priests, and the other, more external, of the people and also the Great Atrium according to 1 Kings 6: 36 and 7: 12; Ezekiel 10: 3; 40: 17, 19, 20, 23, etc., and 44: 17.19, and surrounded with chambers on each side (1 Paralipomenon 28: 12; Ezekiel 40: 17, 44). But those atria were concentric, by which all the gates on each of the sides were equals; from the outer shape to the interior shape was fifty cubits, and between one gate of the exterior atrium and the opposite gate of the interior atrium was an intervening hundred cubits (Ezekiel 40), thus each one of the sides of the exterior atrium by the outside part was five hundred cubits (Ezekiel 42: 20). These atria were not remembered in the Temple of Solomon, except in the small ante-rooms of the cooks and a suburb of fifty cubits in width by which it had been totally surrounded (Ezekiel 45: 2).

At the same time, in the outer wall[519] of the interior atrium was where the Priests were placed. The highest Priests with their Vicars and the great Synedri occupied the most worthy side towards the east. They followed, to the side north and of the south, the Overseer responsible for the services of the Temple and of the Altar, then twenty-four Principles of the Priests, each one of which had their own chamber and, finally, to the sides of the Separate Place. The lower Priests had communally chambers, here they ate the sacrifices and they dressed in the sacred vestments. The exterior of the Great Atrium, the surrounds, were kept for all of the people and here they consumed the sacrifices.

Upon Babylon destroying this Temple, it was built again by Zerubbabel as a Temple with almost the same foundation, which were very similar but with less grandeur and together with an interior atrium that was necessary for the services of the Temple. But with the most external atrium, certainly the richest one and the most extensive one, destined to all the tribes, could not be reconstructed easily neither was it necessary to be built in this place for the few that remained of the two surviving tribes. To the east side of the Atrium of the Priests, a new atrium that was sufficient for the people that had returned from Babylon. But the foundation were reduced in size, the most elderly men that had known the front Temple cried vehemently (Ezra 3: 12). But they did not cry for the appalling appearance of the buildings [because it had not been built yet].

The lamenting had started the moment that the foundation had begun. They regretted that from the measurements of the new sanctuary that it was so narrow and they had replaced the place of the spacious atrium. Although the sanctuary of Zerubbabel compared in its dimensions with the front one, there appears to be nothing to see (Haggai 2: 3). Certainly the Great Atrium was never reconstructed so that the Jews could use it. Indeed Zerubbabel, who was left to build what was necessary for the Jews to worship in, was denied the right to build a sumptuous atrium for the people. Cyrus[520] did not want another extensive one built, therefore there was a decree that the House of God was built with "the height thereof threescore cubits, and the breadth thereof threescore cubits; with three rows of great stones, and a row of new timber (Ezra 6: 3.4)". It is clear that by the three rows of stones and the row of wood it is understood to be the environment of the interior atrium according to 1 Kings 6: 36, where the atrium was thus built as described. Cyrus therefore ordered the building of the Temple and the internal atrium and nothing in addition.

And in this manner the new atrium was built with a simple wall, it was hardly worthy because Cyrus mentioned it in the proclamation so concisely. This sanctuary was maintained without a large atrium until the times of Alexander the Great and subsequently, according to the writings of Hecataeus,[521] a pagan writer of those times. He describes the city of Jerusalem and the Sanctuary with its outer wall of the interior atrium, with its gates, and more of the interior, but down from the outer wall of the outside of the atrium he does not say a word. Although after it was reconstructed that outer wall was more splendid and all men looked to see great things. Thus, the Jews built this in a subsequent epoch, surely under Simon, son of Onias, Pontifice Maximum, who was call Simeon the Just.[522] "For its time he fortified the Temple and it was built the same with the height of double (colonnade), the high fortification of the outer wall of the Temple (Ecclesiastes 50: 1.2)".[523] At last, Herod and the successors completed the work with more sumptuous buildings.

In this Temple, the males of Israel were admitted into the eastern border of the Atrium of the Priests and because of this the border it was called the Atrium of Israel. To the new Atrium of the Women that Zerubbabel had placed on the site of the Great Atrium and that was called the Atrium of the Women, there they entered as many women as males. And those who entered into the large atrium also included the Gentiles. Also the floor of that atrium, that was placed there since the time of the captivity and was out of the limits of the new sanctuary and had no protection of an outer wall of buildings (they had protected it freely after with the outer wall under Simon the Just or better still under the Princes of the Maccabaeorum), they persisted in treading on the soil the same way as before, consolidating the oblivion of the rite of antiquity. It is thought that Solomon had founded the Temple for the use of the idolater, a richer and greater atrium than the one that was used by his own people or that for some reason it had served for the worship of the idolisers, of the children of the ruined, those who God rejects, who were prohibited by the Jews (Acts 10: 28 and 11: 3; Galatians 2: 12) and whose excessive number was despised in Judea in that time. It takes Judea and the pagan domination of the Gentiles to open these entrances to the Sacred Place. For that reason this atrium was called the Atrium of the Gentiles and, by the prophecy of Isaiah (2: 2) which was understood incorrectly as the Mountain of the House; there, as in a profane place, the town traded. And the name of "Sacred" was appointed to the interior of the sanctuary, as if (to say) the exterior atrium could not have been sacred. God, predicting all these things, he corrected them through the prophet Ezekiel.

In the description of the Temple, the Atrium of the Women is omitted and he only speaks of the two atria that are described as having a wall of five hundred cubits of length and width by the outside part "divided the space between the sanctuary and the profane place (Ezekiel. 42: 20)", and again it says: "This is the law of the House, in it higher of the Mountain; all the limit of this is in the outer wall of the Holy of the Holies. This is the law of the House (43: 12)". Finally, it places the Priests in the interior atrium and all of the people are outside (44: 19 and 46: 20, 24), and it does not permit the entry of the Principles and the people to the eastern margin of the Atrium of the Priests, but sends them to do their worship next to the threshold of the interior of the eastern gate (46: 2, 3). And the Jews are blamed for having admitted the Gentile into the sanctuary: "This says the Lord God: that they suffice you already all your crimes even to the house of Israel, because you have induced the alien children, uncircumcised of heart and uncircumcised in flesh, to be in my Sanctuary and they pollute my House, and in the meantime you offer my bread, fat, and blood, and they dissolved my pact with all your abominations. And you do not conserve the custody of my Sanctuary, but you have put guards of my custody in my sanctuary for yourselves. Thus the Lord Jehovah says: No son of a foreigner that is uncircumcised of heart and uncircumcised in the flesh that lives in the middle of the children of Israel will enter my Sanctuary (Ezekiel 44: 7–9)". These things happened in the fourteen years after the fall of the first Temple. This refers also to what had been said by Isaiah to not live as the Gentile idolaters, but of the Jews' and proselytes' future he said: "My house will be called House of prayer for all the people". Christ applied these words to the most exterior atrium and he seriously reproached the Jews for the desecration of this, as strongly as they were allowed in those times. Thus, the Gentile idolaters, whose prayers were abominable, were not expelled by him so that it did not seem that his King of the Jew's authority was exercised untimely against the Romans. Consequently, all the people excluding the Gentile, should be placed there and they were not allowed admittance to the interior atrium and were not allowed to enter through the gate of the south or of the north to do sacrifices, or when they went to the chambers of the teachers (Jeremiah 36: 10).

Dimensions of the Gate on Each Side of the Atrium, as the Shape was Illustrated in Ezekiel 40

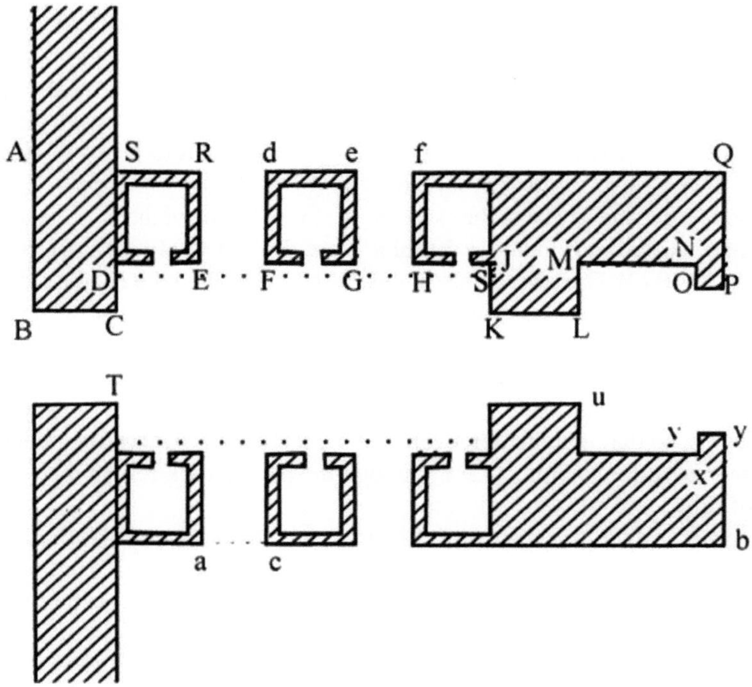

Fig. 1 Image of the dimensions of the gate of the Atrium described in Ezekiel 40[524] (Drawn by the author from Isaac Newton, *Introduction to the Lexicon of the Prophets, Part Two: About the Appearance of the Jewish Temple* (Babson Ms 0434) (unpublished manuscript, Babson College) fol. 8r.)

Verse 5: The Angel measured the width of the external wall **AS** with a reed of six cubits and the height of this with one reed.

Verse 6: Later, the width of the threshold of the door **BC** with a reed and the other width of the threshold **CT** with a reed.

Verse 7: Then three equal chambers that were destined to be for the Gatekeepers, also measured with a reed the lengths **DE**, **FG**, **HJ** and the width **ER** was a single one reed. And the spaces between the chambers **EF**, **GH** each one was of five cubits, and the threshold of the gate **KL** was one reed.

Verse 8: And the other width of this threshold **Lu** was of one reed.

Verse 9: And the vestibule of the gate **MN** was of eight cubits and the front of this **OP** was of two cubits.

Verse 11: And the width of the external threshold **OY** was of ten cubits and the width of the door **NX** was of thirteen cubits. From here the Angel measured all the things with order, and he measured them when in reverse order.

Verse 12: And the width of the six steps of the border of the chambers was one cubit.

Verse 13: And the width of the gate to the wall of the chamber to the wall of the chamber inclusive, advancing according to the line that goes from the entrance of the gate **ac** to the entrance **Rd**, without doubt, **aR** was of twenty-five cubits.

Verse 14 and 16: And the facades of the interior of the gate were through a space of sixty cubits to the facade of the atrium **OP** around the borders to the bend of the gate. That space is: **BC** (6) + **CD** (3 1/2) + **DJ**(28) + **JK** (3 1/2) + **KL** (6) + **LM** (3 1/2) + **MN** (8) + **NO** (1 ½) = 60

Verse 15: And the facade of the exterior gate to the facade of the most interior gate, that is to say, of **B** to **P**, was 50 cubit; since, **BC** (6) + **DJ** (28) + **KL** (6) + **MN** (8) + **OP** (2). Once these measurements had been done he conducted Ezekiel inside the door towards the atrium and saw the thirty chambers from every side of the pavement.

Verse 17 and 18: To the sides of the gate was the lower pavement **BH** (Fig. 2a); in this there were five chambers and thus there were six pavements, so that in total there were thirty chambers. And these chambers were supported by columns (42: 6), so that the people took refuge there in times of rain.

Verse 19: And the Angel measured the most minimum distance **BC** (Fig. 2a), of a hundred cubits, of the two eastern gates of either of the atriums. Here, instead of "and toward the north", The *Septuaginta* reads: "And conducted me toward the north".

Verse 20, 21, 22, 23: The gate of the exterior northern atrium **F** is of fifty cubits long, twenty-five wide and in all ways was similar and equal to the eastern gate **A** (Fig. 2a). And the most minimum distance **FG** (Fig. 2a) of the western doors of either atrium was of a hundred cubits.

Verse 24, 25, 26, 27: The southern gate was similar and equal to the others, and stands apart from the opposite gate to the southern of the interior atrium a hundred cubits. And to these three gates of the exterior atrium ascend though seven steps.

Verse 28: And subsequently, to verse 37, the northern, eastern, and southern doors of the interior atrium are similar and equal to the gates of the exterior atrium, and to these three atriums ascend through eight steps.

Verse 30: Lacking in *Septuaginta* and is corrupt in the repetition of the last period of the preceding verse, where it is put evidently as "five" instead of "twenty-five".

Verse 39, 40, 41, 42: Just in the northern gate in front of it, there are eight writing tables of stone for the use of the sacrifices.

Verse 44, 45, 46: Next to this northern gate was the atrium **RμmN** (Fig. 2b), and in the room **Φ** destined for the Priests that took charge of the custody of the Temple, and a similar room there were next to the southern gate for the Priests that had the custody of the altar. These rooms faced towards the most interior atrium.

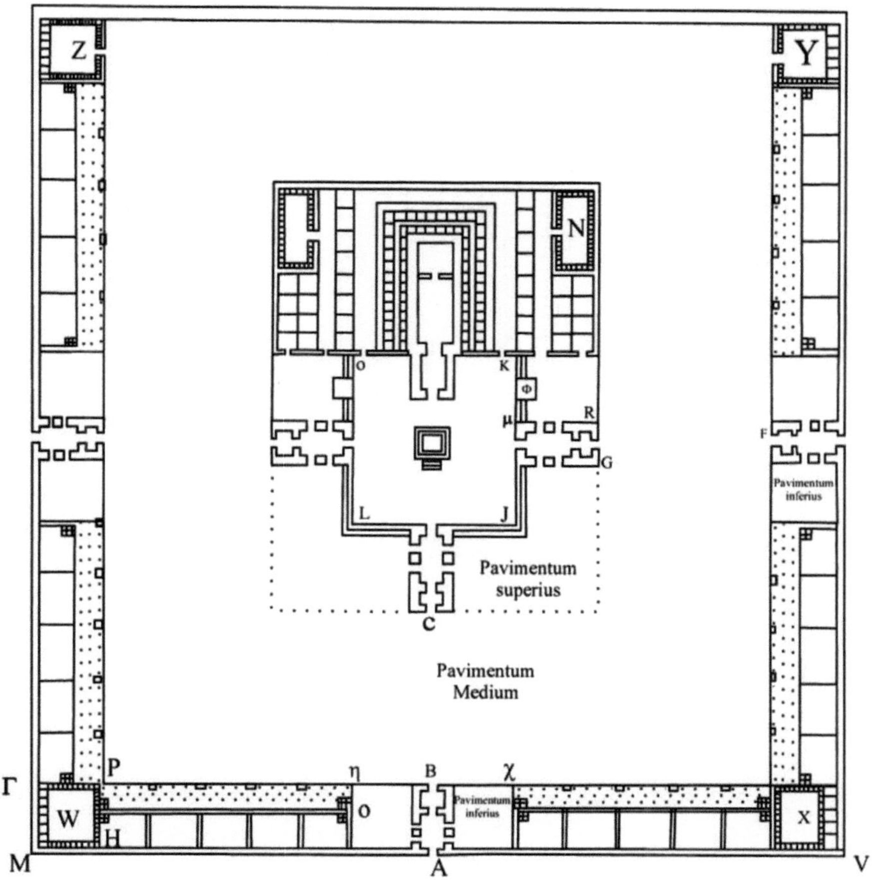

Fig. 2a Floor plan of the Temple of Solomon's precinct[525] (Drawn by the author from Isaac Newton, *Introduction to the Lexicon of the Prophets, Part Two: About the Appearance of the Jewish Temple* (Babson Ms 0434) (unpublished manuscript, Babson College) fol. 9r.)

Verse 47: The length **KJ** of the interior atrium **JKOL** was of a hundred cubits, the width **JL** was the same. Thus, there were two concentric atria.

Verse 48 and 49: **pr** = 5 cubits; **pq** = 6 cubits; **st** = 11cubits (or according to *Septuaginta* = 12 cubits); **sv** = 20 cubits. And ascended to the vestibule of the Temple were ten steps.[527]

Chapter 41, Verse 1 and 2: **wx** = 6; **xσ** = 10; **ya** = 20; **yz** = 40 cubits.

Verse 3 and 4: **bd** = 2 cubits; **bc** = 6 cubits; **bc** + the cavity of wall that covers the entrance = 7. **eg** = **ef** = 20 cubits.

Verse 5 and 6: The width of the wall of the Temple to the bottom of the first chamber was six cubits, jointly with the following width of the chamber from the side was 4 cubits, which gives a width **ab** = 10 cubits. In this way there are thirty chambers or rooms around the Temple and was three floors in height, that is to say, in total

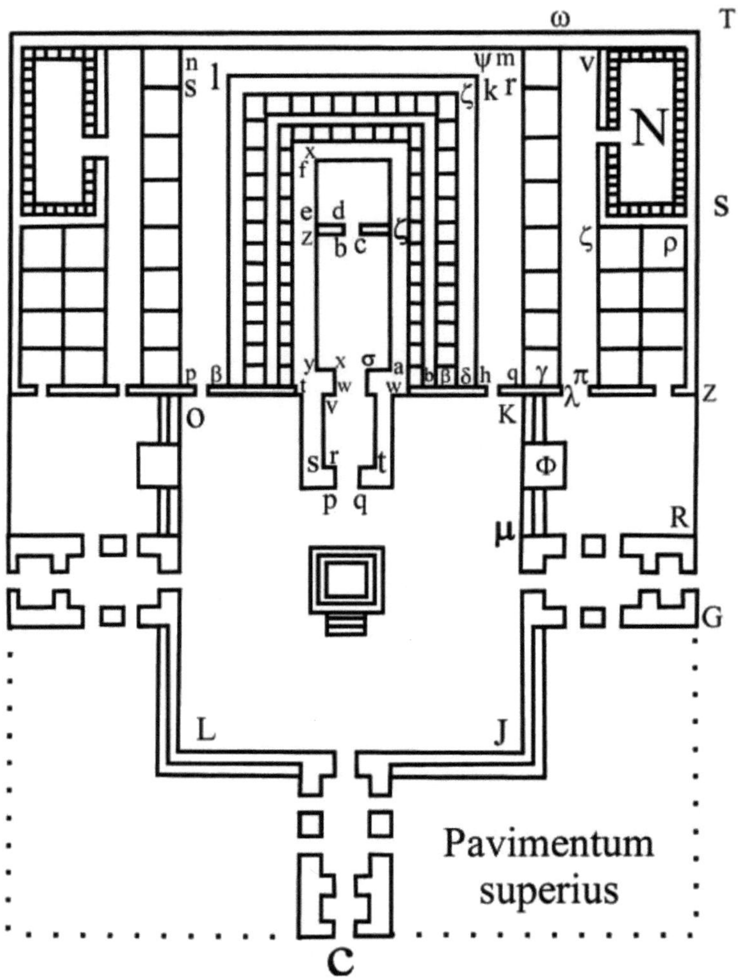

Fig. 2b The inner courts of the Temple of Solomon[526] (enlargement of Fig. 2a.)

ninety rooms (see Josephus 18 Jewish Antiquities c. 3 and the highest rooms were wider than the lowest, yielding the wall of the Temple in three rows of a cubit from every side (see 1 Kings 6: 5.6) so that the first chamber was of five cubit of width (1 Kings. 6: 6), the second of six, the third of seven.

Verse 8: And for all the height of the House the bottom of the rooms was of six cubits long, so that all the length of the thirty rooms measured in total a hundred and eighty cubits,[528] in agreement with the perimeter of the Temple.

Verse 9: Also on the other side of the temple are rooms which are similar and equal to the first **βζ**, from the outside of the wall of the Temple **δζ** is five cubits wide. And the remaining space was to be between the side storerooms of the Temple.

Verse 10: And following the line that is the layout between the chambers there were from all sides twenty cubits, that is to say measuring the width **wh** of all the structure from the wall of the Temple.[529] This measurement is expressed by Ezekiel somewhat obscurely, but is to be compared with a similar phrase to one that he used before in 40: 13.

Verse 11: And the entrances of the rooms were towards the place that was left **hm**, from where they faced each other. There was an entrance towards the south, the other towards the north. And the space that was left **kmnl** was five cubits on all sides.

Verse 12: And the width **lk** of the building **βhkl**, that was in the back[530] of the Separate Place **βγ** towards the western limit, was certainly of seventy cubits, and the wall of the building **δζ** was of five cubits in thickness from every side, and the length of this **hk** was of ninety cubits, was in agreement with the width and the length of the Temple, and from here thence of the buildings. Thus, the width of the Temple (20) + 2 **ab** (20) + 2 **βh** (30) = 70. And the internal length (62) + counting on the wall and the building of the west (25) and the wall of the east the three protrusions of a cubit wide diminished + (3) = 90. Thus, one must remove those protrusions so that the highest building itself will not be higher than the Temple. Those three cubits of the protrusion correspond with the three cubits of thickness of the wall that separated the internal atrium from the Separate Place.

Verse 13: Thus the House with a length of a hundred cubits was measured so that it computes thus: **pr** with the decoration in the shape of the gate was of 6 cubits, for it is such a size the thickness of the wall from every side, add **sv** (20) + **wx** (6) + **yz** (40) + **ze** (2) + **ef** (20) + **fx**(6) and the sum will be a hundred cubits. Besides, the Separate Place **kψ** and the building with its walls **kh** was one hundred cubits.

Verse 14: And the width of the facade of the House **tw** and of the Separate Place towards the east **tp** + **wq**, that is to say, all the width of the atrium **pq**, was of 100 cubits.

Verse 15, 16: And the length of the building to the facade of the Separate Place, to the back of the House **lk,** and the spaces of one and another side **ls** + **kr**, that is to say, the total width of the atrium, **sr**, was a hundred cubits. The Temple inside and the vestibules of the atrium, the thresholds, the windows and the corridors from between the storeroom throughout, in triple floor to the threshold of each one, they were totally covered with wood, etc.

Chapter 42 Verse 1: After he had seen and measured what was in the double interior atrium, the Angel brought me to the promenade of the northern gate **μG** towards the exterior atrium **Rλ** and carried me to the room **λ** that was in front of the Separate Place **hqmk** and in front of the building **ωTZπ**[531] that looked towards the north.

Verse 2: In front of the facade of this the length **λω** was of a hundred cubits, in the gate that looked to the north, and the width **γZ** was of fifty cubits. Thus, it was stated in the Hebrew text.

Verse 3: Of the area of the twenty cubits **ωTZπ** that was in the interior atrium and in front of the pavement that was in the exterior atrium, there was found a colonnade **πζ** facing **λω** in three floors. Thus, it was stated in the Hebrew text. The Latin translator has translated very badly as: "The colonnade next to the triple colonnade".

Verse 4: And in front of the rooms was the passage **λv** whose width **λπ** was ten cubits; it travelled through **γω** for a hundred cubits in length. And the entrances of those rooms that in all this length were placed, they looked towards the north.

Verse 5: And the upper rooms were narrower according to the width: because the colonnade was open in the area of the promenade (to the sides of the highest rooms) removed space (in the Hebrew text they excelled) of these, down from the lowest and including from the centre.

Verse 6: In fact, they were built in three floors and they did not have columns as those of the thirty chambers in the exterior atrium. By this, it is on account of the two promenades with uncovered sides that the middle floor, and the top floor of the rooms was chosen necessary to be shortened and was made narrower than the lower one and the middle floors.

Verse 7: And the wall **ST** that was outside next to the rooms, in the direction of the exterior atrium, by the front of the facade of the rooms, was a length of fifty cubits.

Verse 8: Therefore the length of the rooms **Zρζπ** that were next to the exterior atrium was of fifty cubits, and there, just in front of the facade of the Temple, there was in total a hundred cubits **ZT**.

Verse 9: And down from the place of these rooms, the entrance **λπ** was on the eastern side where they entered to these rooms from the exterior atrium.

Verse 10: In the thickness of the wall of the atrium in the direction of the east and in the direction of the south, upon examining, in front of the Separate Place and in front of the building were the rooms.

Verse 11, 12: Throughout they were similar to the rooms of the northern side.

Verse 13, 14: And he said: "The northern rooms and the southern, opposite the Separated Place, were sacred, they were the holy chambers where the Priests shall eat, once finished their service they shall leave their vestments before leaving the exterior atrium of the Temple, to go to the place of the people".

Verse 15: And when the Angel measured the House inside, then he brought me into the eastern gate of the exterior atrium **A** (Fig. 2a) and measured the fourth exterior sides of the perimeter of the wall, each one of them **MV** was of five hundred cubits. The Sanctuary was limited by this wall to the profane place.

Chapter 46, Verse 19: After I had seen the glory of the Lord, he brought me into a place **N,** where the Priests cook the sacrifices and in the corners **W**, **X**, **Y**, **Z**, of the four atria the most far away atrium was where the sacrifices of the people were cooked. Ezekiel had not seen before these small ante-rooms, because they were hidden by five chambers and five on the pavement below that joined in angle **P**, they were concealed, so that the smoke could be not seen in the two atria. Thus, those chambers, together with the colonnade, they occupied all the width of the lower pavement. But they did not occupy all the length of this to the gates because the Angel measured there the width of the exterior wall and, besides, the gates are accessible by side arches. On the other hand, it is added to the space between the one chamber and another part of each door. There remains the necessity to define the optimum symmetry of the buildings. And for which I know, would be optimum symmetry if that space was to correspond to the width of the lower atrium. We suppose, therefore, this space **ηχ** to be of a hundred cubits,[532] **Pη** to be a hundred and fifty cubits, **Ho** the area of the chambers, and **Po** the area of the three colonnades in front of the triple floor of chambers.

Ezekiel Chapter 40 Verses 14 and 16: And then the Angel, upon returning, measured the facades to the inner gate through the space of sixty cubits to the facade of the atrium **OP;** he measured in the contour of the gate, that is to say, advancing not according to a straight line, but through the contour and the inflections of the door **BCDJKLMNO** (Fig. 3). There is in total **BC** (6) + **CD** (3 ½) + **DJ** (28) + **JK**(3 ½) + **KL** (6) + **LM** (3 ½) + **MN**(8)+ **NO** (1 ½) = 60.

Ibid Verse 44: The Hebrew text and the Latin version are especially corrupt. Ezekiel is conducted then to the most internal atrium, and when brought from one place to another, this is the custom in expressing it. Therefore it reads in the *Septuaginta*, "And he brought me to the interior atrium and in the interior atrium there were two rooms; to the back **Φ** (or to the side) of the northern gate looking at the south, the other to the back of the southern gate facing towards the north". As Ezekiel first came to the exterior atrium, to express his observations of the chambers, thus he also did this in the interior atrium. The first room **Φ** was destined for the Priests that kept watch at the guard house, the second for the Priests that watched for the service of the Altar. The words "singers" and the "eastern door" are read already in the Hebrew text but do not appear in *Septuaginta* and this greatly confuses their meaning. Certainly the Levites do

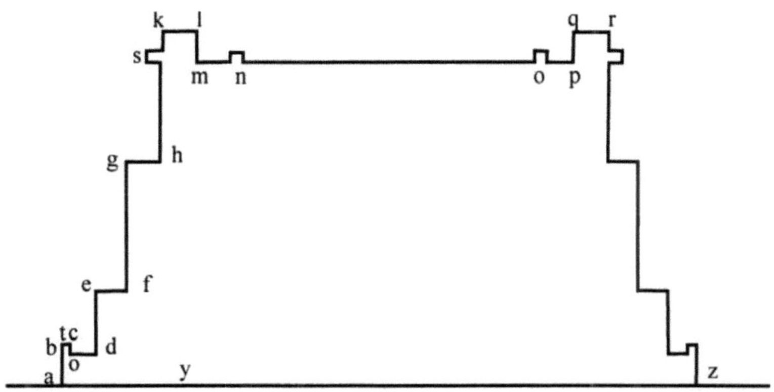

Fig. 3 Contour of the altar[533] (Drawn by the author from Isaac Newton, *Introduction to the Lexicon of the Prophets, Part Two: About the Appearance of the Jewish Temple* (Babson Ms 0434) (unpublished manuscript, Babson College) fol. 13r.)

not have chambers in the Atrium of the Priests, a great deal less in the noblest place than is granted to the Priests.

Ezekiel 41 Verse 6: "And the side rooms, which were in upper positions, were three times thirty and doubly distributed". Thus this translation agrees with *Septuaginta*. That is to say, there were three times thirty next to the Temple and another three times thirty in the opposite side in the upper positions, that is to say, thirty in the bottom, thirty in the middle and thirty on the top. In the Hebrew text is said thus, "And the side, that they were upper positions, were three, and thirty" פעמים (since this way it is necessary to punctuate this word) "twice in succession".

That is to say, the side upper positions, according to the height were three; according to the length were thirty, and this for two times, the thirty of a side and another in coherence with the version of *Septuaginta*. See Ezekiel 41: 16 and 1 Kings 6: 6 upon the three floors and Josephus (*Jewish Antiquities* l. 8, c.3) upon the thirty chambers.

Verse 8: And I saw to the height of the House from all sides (that is to say, by the three floors of storerooms of this one) the foundation of the side storerooms that measured a reed of six cubits. *Septuaginta* puts instead of "foundation" διάστημα. These upper storerooms were wider but not longer that the lower ones, the space of the storeroom length was the height of the entire House.

Verse 9: In Latin it is said thus, "And there was an interior space of the House in the side of the House". But whether that was the interior of the House itself, is not read in the Hebrew text. And *Septuaginta* reads: "And the remaining space was between the sides of the House".

That is to say, it is the space out of the way in the middle of the sides of the house. That "remaining space" also in the Latin version translates very badly, as the "interior of the House;" in the Hebrew text is said תיב חנמ but חנמ does not signify "interior" but "free and empty", the open space occupied between; and next in verse 11 is spoken of the space towards the one that looked at the entrances of the storerooms of each of the sides that, is said, they had five cubits wide from every side around, and nearest is the Hebrew that reads תיב "House", *Septuaginta* translated it for ἀναμέσον and therefore reads ןיב. It therefore reads בין with *Septuaginta* and the sense will be clear. Thus it is translated literally as, "And the empty space that remained was between the sides next to the House". That is to say, what remained beside the buildings was a pavement between the side storerooms that were situated next to the parts of the wall that faced the House. Or more briefly translated in the *Septuaginta* as, "Remained as a space between the House sides".

Verse 10: "And between the chambers that were around the House on all sides the width was of twenty cubits". Though these rooms here are not to be understood as the rooms of the Priests (those were not situated around the House on all sides nor as far as this can be seen were they named as such), but as the side storerooms. And this measurement is not the width of the intermediate space (that was only of fifteen cubits, see Verse 12), but it was the width of the storerooms inclusive, the width of all the outside space of the Temple that occupied the storerooms. Whereof, upon describing subsequently the rooms of the Priests' chambers, these were, on the one hand, in front of the twenty cubits in the interior atrium and, on the other hand, in front of the pavement which was in the exterior atrium. It does not say "in front of the pavement that is in the interior atrium", but "in front of the twenty cubits", appointing not the pavement between the storerooms, but all of the structure of the storerooms with a width of twenty cubits, as if it might have been said "in front of the storerooms". Thus, where it is said that the width between the rooms was of twenty cubits, it is understood as between inclusively, in the same way as is spoken of the width of the gate in 40: 13 from the wall of the chamber to the wall of the measured chamber. It appoints therefore בין not the limits of the intermediate space but situates it in the measurement, and *Septuaginta* translates that correctly as άναμέσον. Through the rooms, cross-sectionally or according to the line that goes between the rooms, it measured twenty cubits.

Verse 15: םיקיחא here it is said of the promenades **krqh** and **lp** (Fig. 2b), and later אחיקים it is said also of three promenades or colonnades from all sides between the three floors of the side storerooms (verse 16), also the promenades are next to the sides of the chambers of the priests (42: 3.5) which I perceive signifies generally "colonnade or promenade without roof or place free of obstacles".

Chapter 42 Verse 3: The Latin version that Villalpando follows, translates colonnades joining a triple colonnade. The Hebrew text has "colonnade against colonnade in three times". *Septuaginta* congruently translates it άντιπρόσωποι στοαί τρισσαί.

Verse 10: I suspect that written there was דדך מידקה "toward the east" instead of דדך מודדה "toward the south" by misconception of the copyist. Therefore *Septuaginta* here has also south.

Ezekiel 40 Verse 14: And then the Angel upon returning measured the facades (perhaps with the seats interposed) to the interior of the door through a space of sixty cubits to the facade of the atrium **OP** (Fig. 4) in the contour of the gate, that is to say, advancing not according to a straight line, but through the contour and the inflections of the gate inside that had already been measured. That is to say, in the perimeter **CD** (3 1/2) + **DE** + **ER** + **dF** + **FG** + **Ge** + **ef** + **fH** + **HJ** (42) + **JK** + **LM** (7) + **MN** (8) + **NO** (11/2) = 62, of which it removes the width of the three entrances (2) and they will remain 60 cubits. **BC**, **KL** and **OP** are excluded exactly the proportion of architectural demand.

Ezekiel 42 Verse 4: "Space of one cubit" in the Hebrew text, but the *Septuaginta* translates "space of a hundred cubits". It reads therefore האמ as "a hundred" where it already reads אחת as "one". Therefore the Angel here measured straight the width of the promenade as ten cubits and the street of this or the space, that is to say, the length of the street etc, as a hundred cubits. It puts therefore דדך "between" the space of the promenade instead of its length.

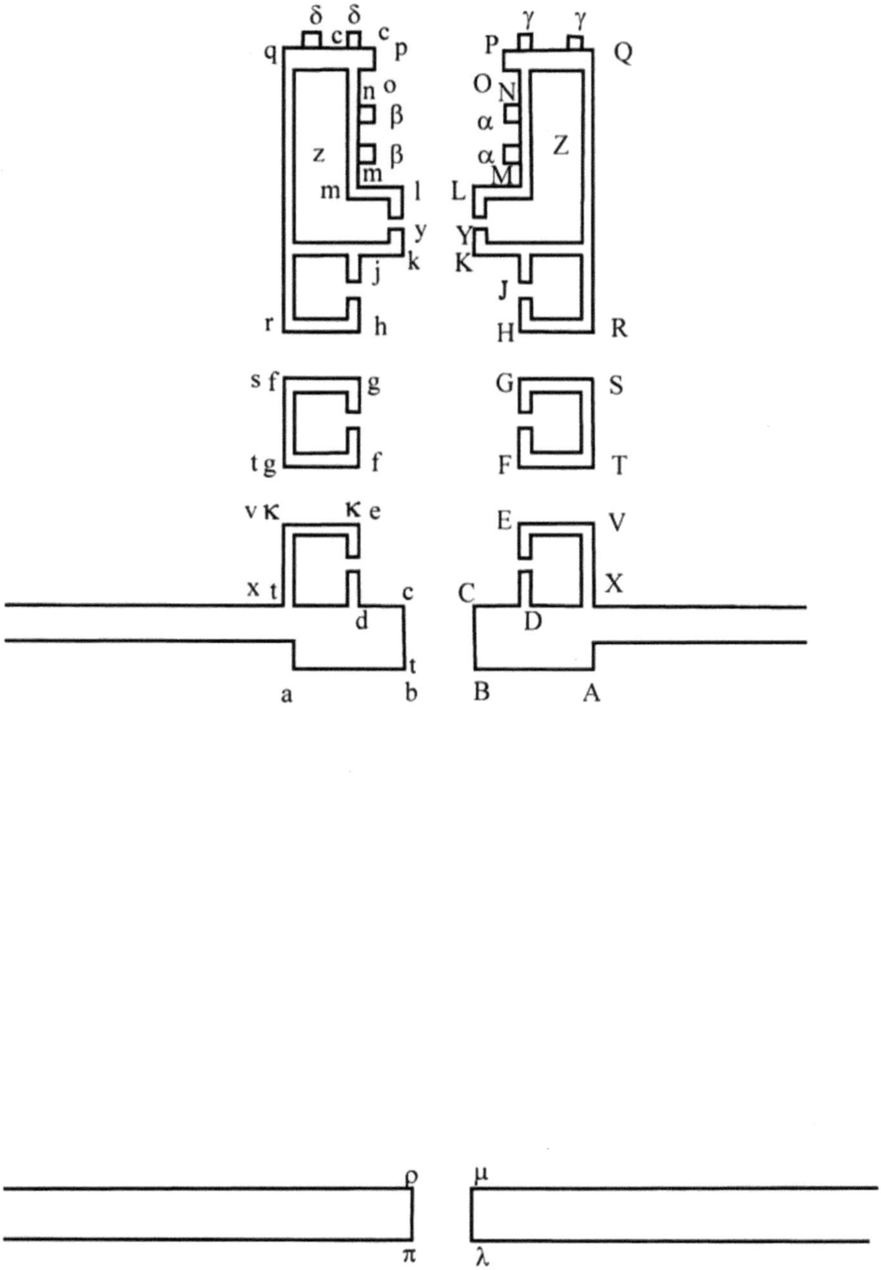

Fig. 4 Ground plan of the Gates[546] (Drawn by the author from Isaac Newton, *Introduction to the Lexicon of the Prophets, Part Two: About the Appearance of the Jewish Temple* (Babson Ms 0434) (unpublished manuscript, Babson College) fol. 42r.)

Commentary

(a) The dimensions of the altar expressed by Ezekiel (Ezekiel 43: 13–17) are thus, its cavity was a cubit **ab**, and a cubit wide **bd**, and its limit **btco** to the edge all around, was given as one palm. And it was the height of the altar [the boundary was used to contain the blood that spilled from the sacrifices so that it might be collected through two small holes to descend towards the abyss, as explain by the experts in Talmud]. And from the cavity of the ground to the lower pedestal was of two cubits and the width was given as one cubit **ef**. And from the smaller pedestal to the greater pedestal was four cubits **fg** and the width was one cubit **gh**. And the same highest Mountain [in Hebrew "Mountain of God," that is to say, the Highest] was of four cubits **hk** and from the cavity of the fire **lq** [in Hebrew "Hunter of God," that is to say, the largest one like the lion. It is a significant large cavity for consuming sacrifices. Here it signifies the great cavity **lmpq** where the sacrifices were consumed] up to the horns of above there was four cubits, and the cavity **lq** was twelve cubits of length for twelve gave the width constituting a square in its four quadrants. And the pedestal [of that cavity] **klqr** was fourteen cubits long and fourteen cubits wide in its four quadrants. And the fringe around **n** or **o** was half a cubit and its depth **lm** was of a cubit in perimeter. To this place Ezekiel went. From where it is deduced that the length and the width of the altar in the ground **az** was of twenty cubits, and the height **ym** of ten cubits, as in the Temple of Solomon. Of the cavity **lmpq** removes the promenade of the Priests in perimeter, the experts of the Talmud say that the width **mn** was a cubit, and it will remain the place of the fire **no** which is ten cubits of length and of width, that is to say of a length and double the width of the one that was built by Moses. Thus, all the altar of Moses was the Place of the Fire, situated inside the promenade of the Priests, and therefore corresponded with the interior space of **no**. The Jews, under the Second Temple to the times of Alexander the Great and further on, they built an altar of twenty cubits of length and twenty in width, but of a height of ten cubits, as recounted by Hecataeus, writer of those times,[534] but later, upon not understanding the mathematical expression "to carry the length to the width," the words of Ezekiel were interpreted erroneously as if the length and the width of twelve cubits had itself to be measured from the centre of the altar. And thus, adding twelve cubits to the correct dimensions, they built an altar of thirty-two cubits of length and width in the base. Remove those spurious twelve cubits and the altar of these will square enough with our description. Again this is evidence that they had the same erroneous interpretation of the words of Ezekiel (48: 20) where there are twenty-five thousand by twenty-five thousand to mark a square which sides are not of fifty thousand each one, as agreed according to the exposition of the Jews, but only of twenty-five thousand (Ezekiel 45: 3.5.6 and 48: 9.13.15). Thus it is said also that the sanctuary, whose length and width was of five hundred cubits, was of five hundred by five hundred (Ezekiel 45: 2).

(b) The Temple of Solomon together with its atria has not been described sufficiently in any place except in the visions of Ezekiel, whose narration is extremely difficult. Certainly, when the Temple was destroyed, there remained buried the print of its foundations in the way the Second Temple was to be built, and it was expected that the buildings would be built up upon the same foundations. We see therefore that the enlightened contribute to the visions of Ezekiel in the contemplation of the Second Temple.

Both Temples were built in an elevation of the Mountain Sion, call Moria. And at the beginning certainly, as said by Josephus (Jewish Wars, Book 6. c. 14), the plain that there was on the top was barely sufficient to situate the Temple for lack of any free space, because there were slopes on all sides. But upon surrounding the hill to the east, to the south and to the west with an immense wall, upon filling the hollow that there was between the wall and the hill, and upon digging into the rock towards the north, they had learnt from the old foundations how great was the space that covered the perimeter square of all the sanctuary. The royal work of Solomon was described by Josephus, who had seen the place, affirming sometimes that this wall had throughout been about three hundred cubits (that is to say, two hundred sacred cubits), but in the eastern side four hundred cubits high (that is to say, two hundred and sixty-six and two-thirds sacred cubits). But it is understood here and in the following as Roman cubits, or more accurately four palms, approximate to the Roman, as will be shown later (Mishna, De Sanedrin Chapter II). And to this wall, the experts in Talmud say, by the interior part there was a step from every side attached, this step was where the people could sit down under a constructed ceiling, which protected them from the rain and the strength of the sun. The banks were situated along a gallery of fifty cubits in width around the Temple (Josephus, Jewish Antiquities Book 15. c. 14). Later, more to the interior, continued Josephus, it surrounded the same high part of another wall of stone, whose east side has all along a double colonnade that looked towards the door of the Temple situated in front in the area of the middle. This colonnade was built by the ancestors of the kings. But this environment and what is more to the interior is described briefly by Philo (Book 2 *Upon the Monarchy*), who had also seen the place. The most external environment of the Temple is broad and open and fortified by a structure of four elegant colonnades; each one consists of a double floor, built with marble, and are very ornamental thanks to the ingenuity of the artists and to the work of the architects, but [thus says the Greek] the interior parts are more modest and they show more austerity in this aspect. In the middle was the Temple which it is impossible to describe with words, only according to what it is possible to imagine can those that see it build it from conjecture, since nobody is permitted into the most interior part to see Him except for the principal of the Priests and even to him He reveals himself only one day of the year. Josephus is in disagreement with Philo, who confirmed that all the colonnades of that exterior environment in the Temple of Solomon (Jewish Antiquities, Book 8. c. 2) are alike in the Second Temple (The Jewish War, Book 6. c. 14) but had been made double, except what Herod built in the southern side that was triple instead of double. In the colonnade of this one (see Jewish Antiquities Book 15. c. 14) there were four rows of columns separated by equal spaces; the fourth [that was the most external one] had a wall of stone. The thickness of each column was that which can be covered by three men joining their arms; the length was of twenty-seven feet with a double moulding underneath. The total number was one hundred and sixty-two with some capitals carved in style Corinthian of a beauty that was a miracle.

With these four rows of columns, continued Josephus, are formed along three intermediate spaces, so that on one and another side are created two equal colonnades, of thirty feet of width, more than fifty in height, with a length of a stadium (stadij), the middle of the colonnade was one and a half times the width and double the height, and they exceeded very extensively that of any other colonnade. The panelled ceiling of these was of carved wood with various figures. Truly, the vault of the central colonnade raised itself very highly elevated upon the architraves of its polished wall of stone, and adorned with distinct columns created the most admirable art to see. Thus it is said by Josephus. Then remove the central colonnade and, for the two remaining that are built to the side, you will have the description of the double colonnade across the circuit of the atrium. Likewise Herod shows its one singular magnificence in the central colonnade because the colonnade preceding, that was visible to them situated in the atrium, should remain in agreement with the other colonnade in the perimeter. But Josephus puts it in another place (The Jewish Wars, Book 6. c. 14), since the internal width of the double colonnade was thirty cubits, that is to say, twenty sacred cubits. One must imagine, since we will not see, that upon the colonnaded rooms had been built, because this was done in the interior atrium of the buildings. However, according to Philo, he considered that they were lower. In each one of the sides of the atrium there were paths to the gate, except in the western side where Josephus placed four gates. The Jews testified to there being seven gates. But the constructions of the corners were different from the colonnades and they were corresponding, if I am not mistaken, to the gates situated in the middle. Thus Josephus (The Jewish Wars Book l. 7. c. 19) speaks of the two colonnades, northern and western, he adds: ώv ή συνάπτουσα γωνία της Κηδρωνος φάραγγοσ ὑπερεδόμητο, "The corners that united, one or other corner was built upon the valley of the Cedar, where the height was immense and admirable". Josephus understands therefore, by the corners of the atrium, not the simple confluences of the colonnades, but some buildings of the confluences.

And from here it so happens that when the Jews were taken, under the Roman Emperor Tiro, who burnt the northern colonnade that when the fire advanced to this corner, the fire ceased and affected very little the eastern colonnade, as narrated by Josephus in the same passage. That is to say, those corners were the four small atria (Ezekiel 46: 22) that by their depth and their high walls easily resisted the fire. Wherefrom also Josephus (The Jewish Wars, Book 6. Chapter 15) says that a corner of the fortress founded by the kings of Asamonaeis, built with magnificence for Herod and so-called Antoniae, was in the corner of the atrium that faced towards the city, so that where it was joined with the northern and western colonnade, had upon them descended to one or other; to the east was the colonnade that was called of Solomon (Jonah 10: 23). From where Josephus (The Jewish Wars, Book 1.16. Chapter 14) thought that only this had been founded by Solomon. And this was situated in front of the Temple (Ezekiel. 46: 2.3) where the Jews preferred to worship, which was also frequented by the first Christians upon yielding worship to God (Acts 5: 12; Josephus, Jewish Antiquities, 15 Chapter 14 and 20 Chapter 8). The remaining colonnades Herod and his descendants built as new constructions. The ancient construction of this one survived up to the fall of the Temple, and even though such a point was emphasized more than the others that Simon the Just had founded it, that in that time deserved the name of Solomon (Josephus, Jewish Antiquities, Book 15 Chapter 14) it was called "Royal" that majestic southern colonnade, and the gate of it obtained the name of "beautiful" for its great ornamentation. Thus, the beautiful gate was the one that the people entered into the Temple by (Acts, 3: 2). But they all entered by the southern gate and they left by the northern one, except those ones which had had something sinister happen to them; they had to enter by the northern door and they left by the southern one, this was written by the experts in the Talmud. Josephus takes these colonnades to be the first sacred place or Sacred Precinct. All the space of the outdoors he said (Josephus, The Jewish Wars, Book 6. Chapter 14), were marked with contrasting colours with all types of stones; the place through to the second Sacred Precinct was surrounded by some fortification of stone of three cubits in height and it was excessively pleasing. This precinct was opened to the southern side as to the northern one by means of three doors equidistantly placed, and from the east by means of a large door by which the pure entered with their women, [but the people were gaining access also into that precinct] where at equal intervals there were columns situated that notified in Greek and Latin writing the laws of purity that meant foreigners should not pass to the holy place. This sacred place is indeed called sanctuary, it rose up fourteen steps counting from the first one, it was square in area and was surrounded by its own wall, whose exterior height, although it was of forty cubits [upon the plan of the Great Atrium], nevertheless was hidden by the steps [to everything along of the south, the east and the north]; the interior height was twenty-five cubits. For in that place on account of the steps set up on high, the interior was not totally separate but was concealed by being covered over. After the fourteen steps there was a flat space to the wall of ten cubits [and that was like the fourteen steps]. From here, another five steps lead to the gates from the north and on the south were eight gates, four on each side [he thinks that the three most eastern of the gates corresponded to the surrounding fortified precinct from all sides]; two were in the east, because it was necessary. In effect, when a proper place separated with a wall [around] in this area was destined for the worship of the women, also it became necessary to have a second gate [in addition to the main and first door situated up in the side of the Sacred Precinct]. This could be found in front of the first area [in the area between that and the eastern gate of the surrounding place]. And from the other places, was the midday gate and another was the northern one, by the one that was entered [going from the space from between the walls situated between the place of the women and the upper square of the Sacred Precinct and since thence] to the place where the women were. Therefore by other doors they did not permit the women to pass [because these gates lead to the place of the Priests where it was not permitted for the people to enter freely]. But neither by their own gate [only] was permitted to cross to the wall that was interposed [upon going towards where the women were]. Therefore only the King or Prince could enter by that gate, the others that came after and had to enter by east of the surrounded place, they deviated from there to the side [going to the gate of the north and the gate of the south]. Therefore that place was likewise open to the native people and to the visitors [as many men as women] that come for religious motives. The western part [of the Sacred Precinct from the latter part of the Temple] did not have any gates but there was a continuous wall that had been built. But between the gates [to the sides of the upper Sacred Precinct] some colonnades were built from the wall towards inside, in front of the rooms [where they were situated between the colonnades and the wall], they were supported by some large and beautiful columns. They were not double as below the exterior atrium but single and besides their magnitude, in no way were they different from below. Nine of the gates [situated precisely in this upper Sacred Precinct] were covered completely with gold and silver, and likewise the doorpost and the lintels, but the tenth gate of the Sacred Precinct [that is to say, the only one that went in the wall of the Atrium of the Women], of bronze Corinthian, which was highlighted covered with a great deal of silver and adorned with gold. And the two leaves of each door were of thirty cubits high and each one was also fifteen cubits wide. And after the entrance, being expanded towards the interior [that is to say, advancing from the wall in which was the entrance towards the interior], they had rooms like towers, to either side, each one thirty cubits in width and forty cubits in height [built upon the colonnade and the rooms]. To each one of these [built inside the two doors] (Josephus, *The Jewish Wars*, b. l.7 c. 13), were two columns of twelve cubits they were supporting the circuit [besides two semi-columns to both sides of the gates that, together with the complete columns, formed three inter-columns under each one of the rooms]. And the size of the other nine doors were equal but the one that was opened upon the Corinthian gate in the highest place of the side of the Atrium of the Women, from the east, in front of the gate of the Temple [the first one between the two of the east], was a great deal larger. Therefore it was fifty cubits of height with leaves of forty cubits and its decoration was richer, therefore it was covered with more silver and gold. This revetment[535] of gold and silver of the nine doors had been provided by Alexander, the father of Tiberii.[536] Fifteen steps [semicircular, according to the experts in Talmud,] they were going up to the great gate from the wall that was separating the women, since they were smaller than those five steps [of half a cubit, according to the experts in Talmud,] that lead to the other eight gates. And the Temple situated in the middle [not in the centre but in the halfway line], was called the Holy of Holies, it rose from twelve steps. The height and the width of this in the facade was of a hundred cubits each one, but behind (Ezra 6: 3) was narrower by forty cubits. It emerged as a kind of shoulder from all sides that was twenty cubits [to such an extent was expanded the vestibule]. But although the Temple was separated interiorly [to a height of sixty cubits] in floors [having built an attic in the area of above], only the first chamber [be it understood as a vestibule] was opened to the totality of the height and raised itself to ninety cubits with a length [interior] of forty cubits and twenty in a transverse line. Near the sides of the lower part of the temple [that was raised to the upper floor] there were many accessible chambers with triple floors placed some upon others, and they were opened to each other's side [of the Temple] entrances through the gate [of the Temple, according to the experts of the Talmud, in the middle of this colonnade, between the gates of the Temple situated to both sides of the wall]. But the highest part did not have these chambers and in this measurement was narrower, but forty cubits high and not so decorated as the part below. It is deduced therefore that the total height was of a hundred cubits, upon adding the sixty cubits of the ground floor. The altar in front of the Temple was fifteen cubits in height, fifty in length and width and was square, had in its corners which were projections as of horns, and an enough slope that inclined to the south. It had not been built with iron nor at any time was iron ever touched. There was surrounding the Temple and the altar a precious fortification of polished stone of a cubit in height that separated the people [in the direction of the east] from the Priests. The entire city was maintained far away from the ones that suffered gonorrea, that is to say, the ones that cannot retain the semen, and the lepers; also the entrance was prohibited to the menstruating women and it was not even permitted to the free women of impurity to pass the limit as mentioned before. To the males that had not been chaste and to those of the Priests that were being purified was prohibited them access to the interior atrium [this was interposed by the fortification]. This point Josephus repeats in another place (Josephus, *The Life Against Apion*, b. 1.6). The Temple had four colonnades [or atriums that were equipped with colonnades] that surrounded it, and each one of them had its own protection according to the law and thus to all was permitted to enter into the outside, also to the foreigners [by this was called Atrium of the Peoples]; only were the women in menstruation prohibited to pass. To the second colonnade [thus called the balcony that was constructed like the colonnade] all for the Jews and their wives if they were clean of every impurity entered here. To the third, the Jewish males that were maintained pure and purified entered. To the fourth, the Priests put on their sacred vestments. To the most interior part, only the principals of the Priests could enter. The experts in Talmud write testimonies that agreed with this (they consulted the *Middoth* treaty which has been explained by Constantinus L'Empereur. Likewise Cappel and Arias Montano demonstrated that they departed from rabbinical material) saying that the Great Atrium (to the one that was called Mountain of the House) had been square, of five hundred cubits in length and in width by the outside and covered with wood, with a colonnade in front of the colonnade, that is to say surrounded by a double encircling colonnade: the interior part was surrounded, in the first place, by a wall with tiles that measured two Jewish vulgate cubits high called גרוס Soreg,[537] later by a solid interior wall called ליח Chajil,[538] whose height in the east was of ten cubits, but by the other sides was greater; finally, by the space from between the walls of ten cubits wide and where the gates were found and the constructions of the rooms. The experts in Talmud who had not seen the place, confuse the two eastern gates describing only one and assigning this one the conditions of both: enumerate three southern gates that proceed from the west: the Gate of the Ardor, the Gate of the Offering and the Gate of the Water, and likewise three northern: the Gate of the Fire, the Gate of the Offering and the Gate of the Projection. R. Jose[539] adds two more gates to the west, in the southern side called Upper Gate, the other in the northern side called Gate of Jechoniae. In the same way the experts in the Talmud count six rooms in the interior atrium. One must place each one of them between the two gates and one must add more to the outside, another two towards the west, as was done with the gates. Therefore according to Josephus he is clear that there were two rooms in the corners of the atrium towards the west. The Gate of the Projection had an attic, and in it the Priests were doing watch in the upper area and the lower area the Levites had the entrance that was opened towards the wall Chajil. And thus also the House of the Ardor was vaulted over and it was large, and had four lower chambers: two in the holy place, that is to say in the area of the colonnade, where they were open towards the interior atrium; two in the profane place, that is to say on the outside, in the area of the rooms that were closed from the atrium and by this name situated out of the atrium, that is to say, in a comparatively profane place. And these two were opened towards the profane place, that is to say, towards the space that existed between the walls. Likewise, the room that was called the Square Stone was situated in part in the holy place, that is to say, upon the colonnade, and in part in the profane place, that is to say, upon a room, was opened towards the profane place or a space between the walls and was a great and a sacred place. Therefore this is where the great Synedrium[540] of seventy elders sat. You can imagine that the other gates and rooms were similar to them described so that a uniform atrium was formed. But in the eastern gate there were only two lower chambers, the Room of End, responsible for the vestments, on the side of the gate on the right or to the north, and the Room of the Utensils of the Cooks was to the left. Wherefrom this it is deduced that that side of the atrium was narrower and lacked rooms. In the corners of the Atrium of the Women, in the area outside of the Place of the Rooms, there were four small atria each one forty cubits long.

These had been built by Zerubabel instead of the atria that had been in the corners of the Great Atrium. The width of the Atrium of the Women (between these small atria) the experts of the Talmud established was a hundred and thirty-five sacred cubits; they described the walls of this as being light and smooth, and that instead of the colonnade, a balcony around had been added, so that the women, from above, could see the males worshiping below. Under the Atrium of Israel there were storerooms that were opened to the Atrium of the Women. In these the Levites kept the lyres, the psalteries, the cymbals and the other musical instruments. It arose from the east from the Great Atrium to the Atrium of the Women by twelve steps and from that atrium by fifteen semicircular steps to the Atrium of Israel; it was a hundred and thirty-five sacred cubits long and eleven cubits wide. This atrium was separated from the Atrium of the Priests by an elevation of a cubit, of which the experts in Talmud had supposed that the Atrium of Israel had been higher, but, given that the elevation was the fortification of a cubit that Josephus speaks of, you are able to determine that the areas are of the same height. The Atrium of Israel was, at the same time, the eastern limit of the Atrium of the Priests. In the northern limit, between the fortification and the Gate of the Offering, in front of the altar, was the place where the pellets, that were removed from the victims sacrificed, were hanging from eight columns; there meat was arranged and washed on the tables. By the Gate of the Offering the sacrifices that had to be offered were carried to the northern side of the altar, inside the fortification. There were twenty-four rings of six classes, fixed to the ground, where the victims that had to be offered were tied. Between the rings and the tables there were four cubits, and it is evident that from that intermediate space the fortification had happened because the rabbis (Maimonides, *Treaty upon the divine worship*, Book 7 Chapter 1) exposed the case of someone that was situated out of the interior atrium who, having extended his hand towards the inside, killed the victim or collected the blood of the animal's dead body; they said how and where he stretched out the feet of the animal to kill it outside of the atrium. It is evident that the fortification was open there; this also was necessary to carry quickly the dead victims to the place where they would be on hand for the butcher's shop. But at the entrance of the fortification the entrance of the gate had to correspond and that is why there was a gate that was directly opposite the area of the altar. And to the space that there was between those gates the experts of the Talmud assign a width of a hundred and thirty-five sacred cubits and calculate it thus: the ascent to the altar was thirty cubits; the width of the altar was thirty-two cubits; the space between the altar and the place of the rings was eight cubits; the place of the rings was twenty-four cubits; from there to the tables was forty cubits; the columns were four cubits, and of the columns to the wall of the atrium was eight cubits. The remaining twenty-five cubits are, in part, the space occupied by the columns, in part, the space between the ascent to the altar and the southern wall of the atrium. The most modern rabbis divide these into two parts (Maimonides, *Treaty upon the divine worship*, b. 1 c. 5). But I would divide the altar so that it would be situated in the middle of the atrium, which will happen if twenty and a half cubits are assigned to the distance between the ascent to the altar and the wall, and the four and a half of the remaining columns. And better still, of these four and a half I would assign one and a half cubits to the width of the tables (Ezekiel 40: 42), and the three remaining cubits were for the bases of the columns. Thus, in front of the tables placed in straight line, were situated many other columns, likewise in a straight line, being mutually corresponding and, therefore, they did not occupy more space of the width than the equivalent one to the bases of a column. Thus, the centre of the altar situated in the middle was away from the walls of the gates by sixty-seven and a half cubits, of the fortification around fifty cubits, and by consequence, the atrium inside the fortification was of a hundred cubits wide; this was also confirmed by Hecataeus according to Josephus (Josephus, *The Life Against Apion*, Book 1). Thus the width of the atrium was double that of the Tabernacle of Moses, and, besides, the borders of one and another side were seventeen and half cubits wide. The experts in the Talmud estimate the length of the atrium to be a hundred and eighty-seven sacred cubits and calculate it thus: the place of the procession of the Priests that did not minister between the Atrium of Israel and the place of the altar was eleven cubits; the altar was thirty-two cubits; between the altar and the vestibule of the Temple was twenty-two cubits; the Temple was a hundred cubits, and between the Temple was eleven cubits. These all add to, on the one hand, the east margin or Atrium of Israel of eleven cubits, and the sum of a hundred and eighty-seven cubits will be completed. By another side, from the centre of the altar to the western limit of the atrium there was a hundred and forty-nine cubits, being approximately double the measurement of Moses. But since the other part of the centre of the altar to the fortification was of a cubit broad in the direction of the east, where according to double the proportion of Moses there should be fifty cubits, there are only twenty-seven; this happens because the Jews had shortened that atrium from the eastern part so what remained was more space for the Atrium of the Women. Thence and because of this they did not build the room there and the border of the atrium was a third smaller in part than the other sides. In this same proportion I would also reduce the space between the columns of the eastern colonnade. Subsequently the experts in Talmud thus define the dimensions of the Temple relating to the length: the front wall of the vestibule was five cubits; the vestibule was eleven cubits; the front wall of the Temple was six cubits; the holy place was forty cubits; the curtain was a cubit; the most interior part was twenty cubits; the subsequent wall was six cubits; the width of the western chambers was six cubits, and the wall of the chambers was five cubits. In total it was a hundred cubits. The width from the Temple calculate it thus: the wall of the impluvium was five cubits; the impluvium or basin in the atrium floor to receive rain-water from the roof of the temple was three cubits; the wall of the chambers was five cubits; the width of the chambers was six cubits; the wall of the Temple was six cubits, and from there to the centre of the Temple was ten cubits. In total, thirty-five cubits, a quantity that was double the total width of seventy cubits. Otherwise, I do not see clearly why so thick a wall would have been built to contain the rain. I would prefer to build an elegant fortification of two cubits wide and high only and to add to the chambers the three remaining cubits of width. Thus, all the width of the Temple with its chambers would be of sixty cubits entirely, as confirmed by Josephus and Ezra (6: 3). The experts in Talmud enumerate thirty-eight chambers around the Temple, surely fifteen giving to the south and fifteen to the north, that is to say, five in each one of the floors, and to the west three on the ground floor, three in that of the centre and two in the highest one. From the atrium it was raised up to the vestibule of the Temple by twelve steps.

The entrance of the vestibule was of twenty cubits wide and forty high; the door of the Temple was ten wide and twenty high and had four leaves, two inside and two outside. The outsides were opened towards the interior of the vestibule, the interiors within the Temple. To the sides of the vestibule, to one and another part, there were two small atria. Through the northern side entered the one that was wanted to open the doors of the Temple; from there an opening that existed in the middle of the thickness of the wall went up to another entrance, that of the doorposts of the Temple, and from it towards the space that there was between the exterior and interior leaves and where it was opened to them. For this same opening went to the spiral staircase that was in a corner of the Temple, from where a third entrance leads to the lowest chambers, a fourth to the middle and a fifth one to the highest. Therefore the experts of the Talmud situate the five entrances in the corner of the Temple, which was oriented towards the east and the north. And from there the Temple could be surrounded going by chamber to chamber within each floor by the entrance of the intermediate wall. This approximately was what was described by the experts in the Talmud.

So that we can complete the description of this Temple one must compare between the measurement of Josephus and that of the experts in the Talmud. It is not the place to discuss extensively upon these things. I will say briefly that the experts in the Talmud use the sacred cubit of six palms and that the Jews, instead of the measurements of the foreign peoples, employed measurements of their own under the names of the Gentiles, such as the sacred cubit for the smaller Roman passu; two sacred cubits for the greater passu; one thousand sacred cubits or Berah for the thousand smaller sacred cubit; two thousand sacred cubits or the Sabbath way for the thousand of the greater passu; the measurement of four sacred palms of the cubit of the Greeks, in the same way four hundred cubits of the Greek stade and the length of the run of the horses by the valley up to the Temple, approximately seventy reeds or four hundred sacred cubits, by the greater stadium. And with this stade Josephus described the measurement of the exterior atrium. The experts in Talmud (*Talmud Hierosolymitanus*, Chapter 6 see also Buxtorf, *Lexicon Templi* in מיך and Arias Montano, *De mensuris*) interpret the other stade as being equal to a thousand steps, that is to say two thousand sacred cubits, with the seven and a half stades. And Josephus uses extensively the cubits of this one stade in the description of the Temple writing for the pagans, except in some measurements of the Temple in a strict sense, repeated and remembered in the Sacred Writing, where he thought that it was necessary to preserve the sacred cubit. There will be constancy of this in the comparison of the cubits of Josephus with the sacred cubit of the experts in Talmud in the following table.

	Cubits as cited by Josephus	Cubits reduced to sacred cubits by Josephus	Cubits as cites by the experts of the Talmud
The height of the wall of Chajil	40	26½	
External	25	16 2/3	
Internal			
Difference of steps 19 or 20 that correspond to half a cubit in height	15	10	9½ or 10
Height of the surrounded (wall/fortification/barrier) precinct	3	2	2 Vulgar cubit
Gate – Height	30	20	20
Width	15	10	10
Altar – Height	15	10	10
Width	50	33 1/3	32
Perimeter of the Columns	12	8	8
Internal height of the Temple	60	40	40
Width of the door of the Temple	16	10 2/3	10
Height NE, perhaps but correctly ΛB	32	21 1/3	20
Internal width of the vestibule	20	13 1/3	11
Width of the entrance of the vestibule	25	16 2/3	20
Height of the entrance of the vestibule	70	46 2/3	40

Thus, when the measurements of Josephus are reduced to the sacred measurements evidently they are in agreement with the ones cited by the experts in the Talmud or, better still, approach them. Therefore Josephus, upon writing for the gentiles, was not careful enough with the measurement as is evident and he made mistakes. Once acquainted with them we can see how Josephus measured the Temple.

To the rooms that were between the two gates, Josephus assigns thirty cubits, that is to say, twenty sacred cubits. Often Josephus used rounded numbers. It is better to write twenty-two, since the two columns upon which each room was supported itself cannot be included in a smaller space. The perimeter of these was of eight sacred cubits (this is in agreement with Josephus and the experts for the Talmud). For that reason the diameter of the upright of the columns was of two and six elevenths cubits. In a proportion taken upon this, one was from three to two, the width of the base will be of three and nine elevenths cubits or almost twenty-three palms. We suppose that it is only of twenty-two palms. The intervals of these pedestals, according to the proportions of architecture, should not be less than the pedestals.[541] If these have equal intervals [since that proportion is very simple], the two columns with as many semi-columns and the three inter-columns will occupy a space estimated at twenty-two cubits between the two gates of a room. I would adapt the gate to the space of the entrance so that the room of both wings, with a width of eleven cubits, and the space of entrance with a width of ten cubits, makes the total width of thirty-two cubits, therefore the widths of eleven and twenty-two cubits are reached enough in these buildings, and the doorway of all the gates, except the first eastern, according to testimony of Josephus and the experts in the Talmud, were of ten cubits in width and twenty high. The authenticity of these dimensions is clearly shown by the harmony of the gates and of the rooms with the parts of the atrium. The two Gates of the Offering have a width of thirty-two cubits exactly; they will correspond to the intermediate altar. And likewise the width of the altar perhaps was enlarged so that this corresponds with the gates. The two following rooms, looking to the west, of a width of twenty-two cubits, will correspond with the space of this same width between the altar and the vestibule of the Temple. The two Gates of Fire, the two following rooms and the last gate will occupy all the space of eighty-six cubits in the area of the vestibule and of the Temple that emerge on its side chambers. And the two last rooms, with their western walls of three cubits in thickness, they will fill the remaining space of twenty-five cubits to the western wall of the atrium. The two last rooms face the east, they are twenty-two cubits in width, which corresponds with the space of this same width between the altar and the eastern gate; the western wings of the two last gates are eleven cubits in width, they include the eastern gate and their entrances will include the space that exists between the walls of the same width situated in the middle, and the two eastern rooms that extend beyond, towards the Atrium of the Women. Surely one of these was called the Gate of the Projection because it projected out of the wall of the interior atrium, therefore was in part inside and in part outside. In the same way, there was another gate, called Gate of the Water, which was in part outside. Thus, the priests were the guards in the House of Abtines, built upon it and it is said that they guarded out of the atrium. But the entrances of these gates were out of the interior atrium because through them went the women to their atrium. However, these gates were partly in the interior atrium because they were found between the nine doors of silver of the ones that Josephus separates; the door of bronze situated in the Atrium of the Women he called ἔξω τῆς νέως, out of the Temple, that is to say, out of the Sacred Precinct of the Temple. Thus, the nine remaining gates were in the outer wall of the more interior part of the Sacred Precinct. The Gate of Projection was called the Gate of the Singers by R. Jose. Surely the Levites sang in the two places, one upon the fifteen steps, that is to say, in the space that is found between the walls, between this gate and the Gate of the Water; the other situated next to the fortification, a cubits wide space looking towards the altar. Finally by this, because all of the people entered through these doors both to the Atrium of Israel and to the Atrium of the Women, one must say in general that these had an intermediate place since they were conducted to the space that existed between the walls situated between one and the other atria. We have therefore exactly the position of these doors and, besides the position, also the magnitude of the other doors and of the rooms to the western wall of the atrium. It is agreed without doubt that all these gates had been equals and that they were equidistant from the others, and besides it is deduced from the gates of the opposite fortified place whose distances Josephus described as being equal. But for some other reason, the four doors cannot be situated regularly at equidistance according to the length of the atrium.

Once they mark the boundaries of these things thus, we have at the same time the thickness of the eastern gate of eleven cubits that corresponds to the width of the western wings of the Gate of the Projection and of the Gate of the Water. This I calculate thus: the bases of the columns were twenty-two palms; the width of the promenade was four and a half cubits; the half pedestals were eleven palms; the interwoven wall to the semi-columns was a cubit; the sum eleven cubits; add all of these together with the atrium which is as mentioned before a hundred and eighty-seven cubits, and the total length was two hundred cubits. Thus the Jews maintained the double length of the atrium Tabernacle inside the fortification of a cubit, in the total length of the atrium with the buildings included. Again it will be taken [according to the above reasons exposed] the width of the promenade of the northern colonnade to the width of the promenade of the eastern colonnade is to the proportion of the limits of the adjacent atrium with the columns, that is to say, it was of seventeen and a half to eleven cubits, and the width of the promenade of the colonnade will be possibly near to double the width of the width of the base of the column. All the width of the atrium between the walls of the columns will be of a hundred and sixty and two thirds cubits. It should be added to one and another part a room of almost eleven cubits, including the walls, and the place of the steps of almost eight cubits, including the most external wall. And all the width of the atrium, including the buildings, will be of two hundred cubits in agreement with the description of Ezekiel. Thus the Sacred Precinct, with a length and width of two hundred cubits, will be square and thus that beautiful and visible structure will be reproduced for the ones that pass by, and surrounded by the space from between the walls of the experts in Talmud of ten cubits of width. Josephus confirms clearly that he estimated it in this way. By these things he affirms that "the upper Sacred Precinct was square and encircled by its own wall". He understands therefore that not the entire Sacred Precinct was included, such as the Atrium of the Women as Cappel thought, but that the upper plan of this Sacred Precinct was, up to the one that was ascended by fifteen steps from the Atrium of the

Women and that more strictly was called the Sacred Precinct. The words of this τητράγωνον δὲ ἄνω "squared up" clearly signify the above plan.

At the same time, the wall that Josephus and the experts in the Talmud wanted to surround the Sacred Precinct, that is to say, the wall of Chajil, I would estimate its thickness, on the one hand, because it was the most exterior one and contained inside the leaves of the doors; on the other hand according to Josephus (Josephus, *The Jewish War*, b. 1.7 c. 22), because of the Romans ramming it without interruption for six consecutive days before the walls of the cities would yield without difficulty, but they did not obtain anything at all. Thus, with a thickness of six cubits this wall will be equal to the most exterior wall of the Temple of Ezekiel 40: 5, and all the width of the sanctuary, including the space from between the walls and the wall of Chajil, will be of two hundred and thirty-two cubits in length. The internal width will be reduced to a hundred and thirty-five cubits and half of the remainder will be the length of its gate, forty-eight and a half cubits; if to this you add the ornaments of the facades of both sides, easily there will be completed the length of fifty cubits that Ezekiel assigned to the gates. But since the gates are expanded in this manner, through the centre of all should pass the space between the walls with entrances which, when transversed, are ten cubits wide and twenty cubits high.

With these measurements the width also corresponds to the measurements of the Atrium of the Women. The small angular ante-rooms, that were situated in the place of the small ante-rooms of the corners of the Great Atrium, should corresponded with those both in width as well as in length. And thus, the Atrium of the Women was of a width of one hundred and thirty-five cubits, and the two small ante-rooms are thirty cubits in width (Ezekiel 46: 22), together with the walls of five cubits in width, they will complete a width of two hundred cubits inside the space that exists between the walls, in the same way as in the upper atrium. All the computations of width are confirmed mutually with so large a correspondence.

Coming out of the wings of the gates of the corners was where the experts of the Talmud situated the small ante-rooms, and the length of the Sacred Precinct looking towards the east, if it was begun from the eastern colonnade will be computed thus: space from between the walls was ten cubits; wall of this was a cubit; the small atrium was forty cubits; promenade with its walls between the small atria was nine cubits; another small atrium was forty cubits, and sum was one hundred cubits. The sacred enclosure can be added on the one hand as above, two hundred cubits, and on the other hand, the eastern wall of the Atrium of the Women was two cubits, the space from between the walls of a part and another, ten cubits, and the wall of Chajil, six cubits, and will have all the length of the Sacred Precinct, three hundred and thirty-four cubits. This length is confirmed with a double proof. One of these proofs is that of Hecataeus and Josephus (*Against Apion*, Book 1) who counted that the outer wall of stone of this Temple had a length of almost five plethris. Likewise, Hecataeus speaks of a length of a hundred cubits inside the fortification. Those hundred cubits are sacred cubits and, because Hecataeus, a pagan male that lived in Egypt, had received this measurement from the Jews in sacred cubits he had not taken into account to convert them to the Greeks cubits, upon hearing cubits he thought of the vulgar cubits and, then by simplifying the expression, to plethros. When they are converted to cubits then the length of the outer wall of stone will be of three hundred and thirty-three and a third cubits or, in round numbers, three hundred and thirty-four.

The other argument is that the altar should be in the middle of the whole precinct, since in another time it was in the middle of the Great Atrium and the Jews would not have easily changed the place of this. Also the interior of the Sacred Precinct should be in the middle of the Great Atrium; in its western and eastern limits it has to be equal distant of the sides of that one and with its Gates of the Offerings, it had to look directly towards the gates out in the middle of the remaining sides. The altar was thus in the common centre and this is supported by our description. The altar is in the middle of the interior atrium that ends, on the one hand, in the foyer of the Temple, and on the other hand, in the eastern vestibule; while on the other hand it is in the middle of the whole interior sanctum, being distant from the external wall of Chajil, for both sides are one hundred and sixty-one cubits and, finally, in this way also from the Great Atrium. The Atrium of the Women cannot increase or diminish one single cubit without it ceasing from being concentric. As the space between the altar and the eastern colonnade, of one part, and the space between the altar and the vestibule of the Temple, of the other part, with a width of twenty-two cubits, correspond exactly. Thus the buildings between the interior atrium and the Atrium of the Women, of one part, and the vestibule of the Temple between the interior atrium and the Separate Place, of another external width of twenty-two cubits, correspond mutually. Thus the Atrium of the Women, of one part, and the Separate Place, of another part, with a width of eighty-nine cubits, they will correspond exactly. And thus, the eastern wall of the Atrium of the Women and the space from between the walls and the wall of Chajil, of one part, the western wall of the Separate Place and the space from between the walls and the wall of Chajil, of another part, with the widths of the two of sixteen cubits respectively, they will correspond mutually with accuracy. For that reason it turns out to be evident that the experts in Talmud were wrong upon attributing to the Atrium of the Women one hundred and thirty five cubits into the square. In that measurement one must also include the Atrium of Israel, together with the place of the Procession of the Priests, and together with these the western wall of the Atrium of the Women. Subsequently then, it will be required to add forty-six cubits towards the east in which, however, together with the opposite area of the fortified precinct, they did not also include the Great Atrium.

Under the buildings of the upper atrium they had excavated wells, baths and diverse pantries, as for example under the colonnades to the north: the Room of the Salt, in which they replaced the salt for the offerings; the Room of Hipparvae where they conserved in salt the skins of the sacrifices, and the Room of the Bathe, where they washed the intestines of the sacrifices. These were the store rooms because the roof of Hipparva coincided with the soil of the atrium and from the Room of the Bathe left a stair towards the roof of Hipparva. But leaving to a side these and other similar things, we might see what use were the chambers of above.

In the Room of the Cut Stone sat down the great one Synedrium. This was oriented towards the south in the tribe of Judae because it was situated between the Gate of the Offering and the Gate of the Water. Therefore, the line that passed by the east side of the altar abandoned everything that existed towards the west in the tribe of Benjamin;

for that reason, as the altar was in the centre of everything, it was judged that the greater part of the sanctuary be awarded to the tribe of Benjamin.

To the Supreme Priest were assigned its two precincts, one to the south, the other to the north. To the south was the house of Abtines, to the one that is described as high and situated upon the Gate of the Water, that is to say, upon the western wing, and upon the entrance of this the Priest's guard in the east wing. To the north, the Room of the Companions or Assistants that I would put between the other Gate of the Offering and the Gate of the Projection in front of the Synedrii. That room was, therefore, the most worthy one of all. The most worthy side was that of the north and there the most worthy place was towards the east.

Along with the Supreme Priest, I count his Vicars and sub-vicars. Next in dignity were the overseers who were placed in command of the departments of the Temple and of the altar. Since next to these, they had to be assigned to these, therefore, next chambers in dignity: surely the rooms of each side that were between the Gates of the Offering and of the Fire, together with some chambers of the gates that looked towards the east. The overseers, whose responsibility was to make and to prepare the vestments of the priests, were of the lower level; they had in the sanctuary their own room (Maimonides, *Treaty Upon the Divine Worship*, b. 2, c. 7, 20). How much more space was there for the wise overseer?

There remains the last four rooms that were to be divided fairly between the twenty-four Chief Priests of the Curia so that each room of the double floor was three chambers for the six Leaders. Josephus writes that the rooms were of a height of more than forty cubits each, each one of them was supported by two columns. At that height out of forty-five cubits, that is to say, thirty sacred cubits, the rooms were extensive enough to cover the double floor. To that height of forty cubits one must add, that is to say, the height of the columns that I would estimate as six times the thickness according to the style Doric,[542] that is to say, fourteen and two thirds cubits, and add the bases and the capital to them, approximately eighteen cubits. Thus there will be a totality of the height of almost forty-eight cubits. Accordingly the width of each of the three chambers will correspond to the width between the columns.

For the Priests of lower category and the Levites remain the four last gates, since there are not other places where they can consume the sacrifices and justly distribute to all shared evenly. Likewise, each one of the colonnades of the three treasuries correspond to some other inter-column; complete, the entire number of twenty-four rooms according to the number of the Curias. In the first eastern gate there were ninety-six chests to store the vestments of the priests (Maimonides, *Treaty upon the divine Worship* 1, *Tract.* 2 c. 8), evidently four for each one of the Curias with the name of each curia recorded on their chests. Thus also in the wings of the vestibule of the Temple (See note de Constantinus L'Empereur in *Middoth* Chap. 4, Sect. 7) there were twenty-four small cells where the Priests of the Curias independently stored their own sacred knives. Such was the structure of the interior of the Sacred Precinct.

We may descend now and we may leave for the place of the Grilles towards the atrium of the Gentiles. The experts in the Talmud measured it to be five hundred cubits along each of the sides on the outside; Josephus confirms that between the angles was a stade, that is to say, four hundred cubits. Josephus verifies this computation in other places; he assigns that the Royal colonnade is one stade long, and to the entire perimeter of the Temple and of Antoniae, six stades. The perimeter of Antoniae is of six hundred cubits, a very agreeable square according to the description in Josephus. He adds to the perimeter of the Temple two thousand cubits, and the sum will be two thousand six hundred cubits, two hundred of which were absent when Antoniae coincides with the corner of the Temple. A perimeter of two thousand four hundred cubits remains, that is to say, six stades.

The gates, from its length of fifty cubits, should correspond to the corners and from its width of thirty-two cubits, to the gates of the interior atrium. Thus, between the corners of each colonnade and the gate should be distributed twenty cubits between the columns. Josephus speaks of one hundred and sixty-two columns in the Royal colonnade, in Greek ΡΞΒ, better still ΡΞΗ, that is to say, one hundred and sixty-eight cubits by a small change in the two last letters, since the number should be divisible by eight because there are eight equal series of columns, four on one side and four on the another. There is, therefore, in a series twenty-one columns and twenty inter-columns. This is confirmed thus, on the western side of the atrium: of four hundred cubits by the interior part, remove four gates of one hundred and twenty-eight cubits in total and there will remain two hundred and seventy-two cubits as total length of the colonnades in five intervals. Those intervals will be equals, as in the sides of the interior atrium; an interval will be of fifty-four and two fifth cubits and will occupy five, six or seven between the columns, and the Royal colonnade will occupy seventeen, twenty or twenty-four between the columns of the same magnitude. But seventeen, according to the architectural proportions, will be too few, and twenty-four will be excessive if the columns were estimated to be equal to those of the other atriums, and, in one and another case, are set apart too much by the numbers of Josephus, therefore it should be twenty. According to this proportion, the columns will be less numerous than in the proportion of the eustyli[543] of Vitruvius, but more beautiful; and here, where instead of the architrave there are large blocks of marble that cannot be broken, it does not fit the objections of Vitruvius.[544]

And I would equalize the columns of this atrium to that of the others for which they could be included by three united men of arms, as claimed by Josephus, that is to say, that they were of three fathoms or twelve pagan cubits, that is equal to eight sacred cubits. And by this I would also suppose that they were equal to the colonnades in one and another atrium as for its width, excepting the fact that the middle colonnade that was twice as high than the others, would be also double the width in the space from between the base of the columns. Thus, the width of its bases will be of twenty-two palms, that of the exterior space from between the base, double; that of the space of average, fourfold; that of the entire colonnade, triple, at forty-four cubits. The half of the space of the external columns will occupy the outmost wall. Then add the wall that formed the banks, which Ezekiel describes as being six cubits long and wide, and the width of fifty cubits will be completed in agreement with the dimension of the gate and of the corner with the ones that the colonnade finished for both sides. This calculation, with the testimony, is confirmed at the same time by Josephus, who writes that two equal colonnade exteriors of thirty feet wide were surrounded by the middle one, whose width was one and a half times as much. Thirty Roman feet are approximately fourteen and one fifteenth sacred cubits, as will be evident from the following data. Thirty Attics, in Roman feet, are fourteen and one fifteenth approximately. I rather suspect that Josephus was making free with the computations and utilized the foot instead of the Hebrew measurements, which was half a cubit; by this reason, thirty feet becomes fifteen cubits. And in agreement with this measurement, the width of the external colonnade in our description is of fourteen and two thirds cubits, with the columns included, and that of the central colonnade, with the columns also included, is of twenty-two cubits, and by this, one and a half times as much as the front one exactly.

Remove the central colonnade and there will remain the double colonnade from the three remaining sides of the atrium with a width between the axes of the external columns of twenty-two cubits; between them is the base of these which is eighteen and a third cubits, and between the columns it measures nineteen and five ninth cubits. And this is so confirmed by Josephus, who defines the width of the double colonnade in a round number to be thirty cubits that is to say twenty sacred cubits. Thus, we have the dimensions of the colonnade according to the width. But out of the corners will raise up the entire top part, downwards may remain open to the promenade of the colonnades with the doorway into the angles facing the opposite part[545]; they correspond with the sides of the gates.

Now one must add the heights. Josephus writes that the length of a column made of solid stone was of twenty-five cubits, that is to say, of sixteen and two thirds sacred cubits, and in another place he speaks of twenty-seven feet, that is to say, thirteen and a half sacred cubits. From there I deduce that the length was six times the thickness, according to the Doric style, as in the interior atrium, that is to say, fourteen and a third sacred cubits, although this measurement is between the two first. The bases and the carved capital of a column in Corinthian style will complete the length of almost eighteen and a third cubits, that is to say, double one of the spaces from between the axes of the columns. This was, therefore, the height to the architraves. On the other hand, Josephus writes that the internal height of the colonnade was of more than fifty feet, that is to say, more than twenty-five cubits; if this is out of twenty-seven and a half cubits, that is to say one and a half times as much of the height of the architraves from the foundation, the duplicate height of the central colonnade would be of fifty-five cubits in the interior. There will be added to the thickness of the foundation two cubits, the sacred row of sacred chambers whose internal height would equalize to the width of almost fifteen cubits, the roof of five cubits, and the balcony above going around the three floors to complete the height of eighty cubits. In effect, Josephus writes that the height of this central colonnade was very notable.

The Jews built flat roofs on their houses so that they could walk on them. If they open the doorways from the chambers, which were built to the sides of the two outside colonnades, the height of the roofs of the pavements or only of the middle itself will be fifty-seven cubits. Add the encircling balcony of three cubits and the total height will be of sixty cubits. This is the same height of the remaining colonnades in the extension of the atrium, because the front facades of all should be mutually corresponding.

Finally we advance from the Temple described to the Temple of Solomon. One must reject everything that Zerubbabel and Herod added, or what is irregular, as his Atrium of the Women and the Gates of the Projection and of the Water that is correct and that corresponds to the gates of the other side of the altar, but to those ones no other gate corresponds to in the Great Atrium, the four gates with the one double colonnade in the western side of the Great Atrium where nothing corresponds in the interior of the atrium. Thus there will be three gates as a group in the atrium interior that correspond with these of the exterior atrium, and in the middle there will only be the altar. Instead of the direct gates, where the Priests consumed the sacrifices, suitable chambers would be built for this use in an isolated place. And since it pleased Herod and the Jews to be excessive, as to enlarge the internal height of the Temple and the width of the vestibule, of the altar and of the doors, and to put a threefold colonnade in the southern side of the Great Atrium instead of the double one, those defects would be corrected and they would eliminate the two walls, Chajil, and Soreg, with their superfluous steps; therefore these, which, as I believe, Zerubbabel put at first as limits of his sanctuary, diminish and hinder the big atrium dedicated to the most numerous people. The gates would lead towards the inside to the fortification of a cubit long. And, since the Temple with its chambers measured in another time seventy cubits wide (Ezekiel 41: 12), Cyrus diminished its width by ten cubits (Ezra 6: 3) excluding without doubt the promenade that was between the chambers of five cubits wide and building doors instead of this in the middle walls of the chambers that lead the way forward but the Jews, completing the old measurement of seventy cubits, adding to one side and another a wall to protect it from the rain, that wall may be got rid of and restored to the old width of the chambers including the promenade that divided the greater chambers into groups of two smaller to one and another side so that the width of these responded better to the height of five cubits (1 Kings 6: 10). This diminished the length of each one by a similar proportion so there are thirty in total in each series; for instance Josephus enumerates (*Jewish Antiquities* 1.8 Chap. 2) this for the Temple of Solomon. Once it is corrected all of this thus, we will be able to finally imagine in this Temple the thing that Ezekiel's words illustrated with plans and furbishing, which we offer now as follows.

Ezekiel Chapter 40 Verse 1: In the twenty-fifth year of our exile, at the beginning of the year, in the tenth day of the month, in the fourteenth after the city had been smitten, in the selfsame day, the hand of the Lord was upon me and brought me thither.

Verse 2: In the vision God brought me into the land of Israel, and set me upon a very high mountain, by which was the frame of the city [the building of the Temple with its atriums appeared] in the area in front (on the south).[(a)*]

Verse 3: And he bought me there and I behold, that there was a male whose aspect was as of bronze, in his hand there was a cord of flax and a reed to measure and he was standing in the gate.

Verse 4: And the male spoke to me, "Son of the man, behold with your eyes, hear with your ears and set your heart in everything that I show you, for that it be

---

[*]Superscript letters indicate Newton's notes to this text, see pages 152–156.

shown you have been conducted here, declare everything that you see to the house of Israel".

Verse 5: And behold there was an exterior wall νστξ[547] around the whole of the House and in the hand of the male was a reed to measure of six cubits[(b)] [a large one of a constant magnitude] by a cubit [smaller] and a palm. And it measured the width of the building **λμ** with a reed and the height with a reed.

Verse 6: And he came to the gate that looked towards the east, he ascended its steps[(c)] [seven] and measured the threshold of the gate with the reed its width **BC.**[(d)]

Verse 7: The chamber [destined for the manager of the gates] measured it to be one reed in length by one of width **EV**, [the space or vestibule[(e)]] between the chambers [that leads to the colonnade], five cubits **EF**, [and the second chamber was one reed wide **TF** by one reed long **FG**, the vestibule, five cubits **GH** and the third chamber a reed long **HJ** and wide **HQ**].

Verse 8: He measured the threshold of the gate next to the vestibule of the door[(f)] inside one reed **KL**.

Verse 9: And he measured the vestibule of the gate as being eight cubits **MN** and its posts to be two cubits **OP**. And the vestibule of the gate was that part of the inside,[(g)] [or towards the atrium].

Verse 10: There were three chambers to the east [or towards outside] three on this side and three thence, the measurement of the three was the same and also the measurement of the doorposts of this side and thence.

Verse 11: [From here the Angel advanced measuring the length of each thing, and later returned to measure the width]. And he measured the width of the entry of the gate to be ten cubits **Cc** and the length[548] of the gate was thirteen cubits **Dd.**[(h)]

Verse 12: And there was a space [or steps] in front of the chambers, it was one cubit [to a side] and a space of a cubit to the other [being extended to this point the flat surface of the pavements of the colonnade]. And the chamber was of six cubits on the one hand **EV** and six cubits on the other hand **ev**.

Verse 13: And he measured the gate from the roof of the chamber [outside] to its roof [outside] as twenty-five cubits **Vv** in width, **VT** doorway against the doorway **Vt**.

Verse 14: And he did [in the doorway] the doorposts [**BC, bc, KL, H** of a height] of twenty cubits,[(i)] and next to the doorposts of the atrium[(j)] [**TS, PQ, pq, ts**,] there were gates from every side going around **VT; SR, Pp, rs, tv.**

Verse 15: And from the face of the gate of entrance **Bb** to the face of the vestibule of the interior gate **Pp** was fifty cubits.

Verse 16: And there were windows closed [with grilles] in the chambers, in the doorposts of these **EV, FT, GS** and inside the gate round about. In the same way there were windows in the vestibules **MN, nm** in the interior all around and upon the doorposts **PQ, pq** there were palm trees [one on this side and another on the other side as is said in the verse 26].

Verse 17: He conducted me towards the exterior atrium and here there were chambers and a pavement with columns built around the atrium,[(k)] thirty chambers upon the pavement with columns [ten in each one of the atrium side, five to one side of the gate and five to the other, except in the western side that had not yet been seen].

Verse 18: And the pavement with columns was next to the side walls of the gates **XQ, xq**. In front of the length of the gates was the lower pavement with columns [that was above the other atrium].

Verse 19: And he measured the width from the face of the gate down to the facade of the atrium interior from the exterior; it was a hundred cubits towards the east.

Verse 20: He lead me towards the north and the gate was the one that looked back at the north in the exterior of the atrium.(l) He measured its length and its width.

Verse 21: And there were three chambers on one side and the three on the other, its doorpost and its vestibule were according to the measurement of the first gate. Its length was of fifty cubits and its width of twenty-five cubits. Its windows, its vestibule and its palm tree were according to the measurement of the gate that looked towards the east.

Verse 22: Ascending the seven steps towards this and its vestibule was before them [that is to say, towards the interior].

Verse 23: And the gate of the atrium interior [was] facing [this] gate of the north,(m) as in the gates of the east. And he measured it from gate to gate to be one hundred cubits.

Verse 24: He conducted me towards the south and I was at the gate that looked towards the south, and he measured its chambers,(n) its doorposts and its vestibule according to these measures.

Verse 25: This had windows and its vestibule thereof had windows going around, its length was of fifty cubits and the width of twenty-five cubits.

Verse 26: Its steps were seven and its vestibule was in front of these steps and it had palm trees one on each side, upon the doorposts.

Verse 27: And in front was the door of the interior atrium that looked towards the south. And he measured it to be a hundred cubits from the gate of the exterior atrium to that southern gate.

Verse 28: And he conducted me towards the interior atrium, towards the door of the south, and measured the gate of the south according to the same; it measured one hundred cubits.

Verse 29: And its chambers, its doorposts and its vestibule, there were windows on all sides of the vestibules, it was fifty cubits in length and twenty-five cubits in width.

Verse 30: And there were vestibules around **VvtT** and **SsrR** of twenty-five cubits in length and five cubits in width.(o)

Verse 31: Its vestibule was looking towards the exterior atrium and there were palm trees in its doorposts and there were eight steps.

Verse 32: And he conducted me towards the interior atrium,(p) towards the east and he measured the gate [east] according to the same measurement.

Verse 33: And its chambers, its doorposts and its vestibule were according to these measured. And the windows were on every side of its vestibule; the length [of this] was of fifty cubits and the width was of twenty-five cubits.

Verse 34: And its vestibules were towards the exterior atrium and there were palm trees in its doorposts on both sides.

Verse 35: And he conducted me to the gate of the north and he measured it.

Verse 36: According to the measurements of above, its chambers, its doorposts and its vestibule, and this one had windows all around.[q] The length of this was of fifty cubits and its width of twenty-five cubits.

Verse 37: And its vestibule looked towards the exterior atrium and there were palm trees on its doorposts on both sides.[r] And there were eight the steps leading up to it.

Verse 38: And [the vestibule had] a chamber [illuminated by the upper windows as described], **Z, z**, and its entrance, and it was in the doorpost of the gates **KL**, **H**, where they would wash the burnt offering.

Verse 39: And in the vestibule of the gate there were two tables on the side **αα** and two tables on the other side **ββ** to slaughter the burnt offering on and the sin sacrifice and trespass offering.

Verse 40: And towards the side [of the vestibule], beyond the step [it is higher than] to the entrance of the gate of the north, there were two tables **γγ** and to the other side of the vestibule of the gate there were another two tables **δδ**.

Verse 41: Next to the gate there were four tables to the side and four tables to the other side, eight tables, upon which they slaughtered [the sacrifices].

Verse 42: And the four tables for the burnt offering were hewn of stone of one and a half cubits in length, of one cubit and a half in width and of a height of one cubit where they put the instruments with which they slaughtered the burnt offering and the sacrifice.

Verse 43: And there were hooks [to hang the meat] of a palm [of length] placed in the vestibule outside, around and upon the tables [also was] the meat of the offering.[s]

Verse 44: And outside the interior gate [to the western side] were the chambers of the Principals[t] [the Priest of the Curias] in the interior atrium [to both sides][u] a [room of many chambers][v] was to the eastern side of the gate of the north; the facade of these was towards the south, the other to the side east of the gate of the south[w] [and] its facade towards the north.

Verse 45: And he said unto me: This room,[x] whose facade is towards the south, will be for the Priests that do the guarding of the temple.

Verse 46: And the room whose facade prospect is towards the north is for the Priests that guard the altar. These are the children of Zadoc that are among the children of Levi to the Senor they serve.

Verse 47: And he measured the atrium [interior] to be a square of a length of a hundred cubits and a width of a hundred cubits; and the altar in front of the facade of the Temple.

Verse 48: And he conducted me to the vestibule of the Temple and measured the posts [one and another] of the vestibule, five cubits of one side and five cubits of the other, and the width of the gate, three cubits from one side and three cubits from the other.

Verse 49: And the length of the vestibule was of twenty cubits, its width[y] of eleven and he ascended up ten steps.[z] And there were columns of bronze next to the post, one on the side and one on the other side.

Chapter 41, Verse 1: He brought me into the Temple and measured the posts. Six cubits in width on the one side and six cubits in width on the other, the width of the doorposts[(a′)] and the width of the entrance was ten cubits.

Verse 2: And the side of the entrance door [to the side of the Temple] was of five cubits on one side and five cubits to the other side. And it measured in length, forty cubits, and its width, twenty cubits.

Verse 3: He entered inside and measured the posts of the entrance door, two cubits; the entrance, six cubits and the breadth[(b′)] of the entrance, seven cubits; from here forty cubits, and its width, twenty cubits.

Verse 4: And he measured its length at twenty cubits and its width at twenty cubits, before the facade [or walls] of the Temple he said unto me, "This is the most holy of holies".

Verse 5: And he measured the wall of the House, six cubits [of width] and [the remaining] width [of the room] every side, four cubits for all the parts around the House.

Verse 6: And the side chambers, that were above, there were[(c′)] all along there thirty chambers, two times [so that the entire number would be two times ninety]. And there were some[(d′)] protrusions in the wall of the building up to the rooms throughout so that those were joined (to the wall) leaning on it, but they were not inserted into the wall of the building [when entering the floor].

Verse 7: And there was a space [a place by which to walk] above and thus above up to the chambers. There was, therefore, a space above the chambers that went still upwards about the house. For this reason there was a width [on that it was possible to walk from the wall] along to the House in the area of above and thus it increased from the lowest part to the highest by the middle floor.

Verse 8: And I saw into the highest place[(e′)] round about, the foundations of the side chambers measured six cubits the whole way to the wing [that one would be able to walk around the projection of the walls. The average width of the central rooms was of five cubits, and insert a cubit to decrease the wall that, added to those previously said widths, give a total width interior of five cubits for the lower room, of six cubits for that of the centre and of seven cubits for the widest one, all according to how it is described in the Temple of Solomon (1 Kings 6: 6)].

Verse 9: And the thickness of the wall[(f′)] of the chambers outside was of five cubits. And the remaining space was between [the rooms][(g′)] the sides that were next to the House [on the one hand] and [the other] the chambers [more removed from the side of the House, by another side].

Verse 10: And the width was of twenty cubits around the House from all sides.[(h′)]

Verse 11: And the entrances of the rooms were oriented towards the space that remained an entrance towards the north and another entrance towards the south. And the width of the space that remained was around about five cubits.

Verse 12: And the building [already described] that [in its external facade] was in front of the Separate Place, [situated] in the area of the west, was of seventy cubits in width. The wall of the building was of five cubits in width around and its length was of ninety cubits.

Verse 13: And he measured the length of the House to be one hundred cubits [excluding the chambers and including the columns of bronze]. And the Separate Place, the building and its walls [jointly], he measured them to have a length of a hundred cubits.

Verse 14: The width of the facade of the House and of the Separate Place [to one side and another] towards the east [as a group], was a hundred cubits.

Verse 15: And he measured the length of the building in front of the Separate Place [western] that was to be to the back part of the galleries[i′] from where and thence was a hundred cubits.

Verse 16: The Temple inside, the vestibules of the atrium [and] the thresholds and the slanting windows, as well as the galleries [that existed between the side galleries] around [the Temple], all three floors [of chambers], facing the threshold of each chamber [of one and another side] were covered with wood throughout. From the ground of the Temple up to the windows,[j′] and the windows [which were made of wood], were concealed.

Verse 17: And to [the area] above the entrance, and also to the House, [all] those inside and outside [to the side chambers], and through all the walls around, inside and outside, there was revetment [of wood].

Verse 18: It was decorated with Cherubim and palm trees. The palm trees were between the cherubim, and all the cherubim had two faces.

Verse 19: So that there was a face of man towards the palm tree on the one hand and a face of lion next to the palm tree on the other side, throughout the house on every side.

Verse 20: From the ground to [the area] above the entrance there were Cherubim and engraved palm trees in the wall of the Temple.[k′]

Verse 21: The doorposts of the Temple were squared. And in front of the sanctuary interior there was a sight of figures [of Cherubim, palm trees and square doorposts in front of the sanctuary].

Verse 22: There was an altar of wood which was three cubits in height, its length was of two cubits, its width of two cubits and it had some protrusions. Its base[l′] and its walls were made of wood. He spoke to me saying: "This is the table that is in front of the Lord".

Verse 23: There were two doors to the Temple and two in the most interior part [aside of the wall, the other to the other side of the wall[m′]].

Verse 24: And there were double doors to the entrance, which had two turning leaves in each door [that of the right and that of the left] and had two double doors in the other.

Verse 25: And there were carvings made in them, in the doors [I repeat] of the Temple, Cherubim and palm trees, the same things that were engraved in the walls. And the beams[n′] were made of wood in the facade of the exterior vestibule [and the floor with the frontispiece was situated on the columns of bronze].

Verse 26: And there were windows [with grilles] closed and palm trees to one side and the other in the side of the vestibule, in the side rooms of the House, and in the entrance.

Chapter 42, verse 1. He conducted me from there to the exterior atrium down to the road towards the north and entered the fifteen chambers[(o')] that were in front of the Separate Place and in front of the building [side] that was towards the north.[549]

Verse 2: The length of the front of the façade[(o')] [of these] was of a hundred cubits in the entrances that look towards the north and the width [of the entire place] was of fifty cubits.

Verse 3: Against the twenty [cubits] that were in the interior atrium and upon the pavement with columns, which was for the exterior atrium, there was a gallery against another gallery in the three [floors].

Verse 4: And in front of the chambers there was a promenade of ten cubits in width inside [but] of a length of a hundred cubits.

Verse 5: The entrances of these looked towards the north. And the upper sacred chambers were narrower because the galleries absorbed part of these away for the lowest [I repeat], and down away from the middle of the building.

Verse 6: In effect, they were built in a triple floor and they did not have columns as the columns of the atrium did. Therefore the building was straightened; there was a decrease in the lowest and the middlemost floor from the ground floor.

Verse 7: The length of the opposite exterior wall of the chambers looking towards the exterior atrium, in front of the chambers, was of fifty cubits.

Verse 8: Because the length [of the other] chambers [that were] next to the exterior atrium was of fifty cubits, while in front of the Temple [added another length] there were a hundred cubits.[(p')]

Verse 9: And under the chambers there was an entrance to the east to enter the exterior atrium.

Verse 10: In the width of the wall of the promenade[(q')] [as mentioned before] that looks to the east. And towards the south,[(r')] in front of the Separate Place and in front of the building [side] were chambers.

Verse 11: And there was a promenade in front of these to the appearance of the chambers that looked towards the north, following its length,[(s')] its width, its exits and[(t')] [all] the layout of these.

Verse 12: And as they were the doors of these,[(u')] thus the doors of the chambers that gave to the south. There was a door at the beginning of the promenade, the promenade [I repeat that was] in front of the wall situated to the east of the entrance of these.

Verse 13: And he said to me: "The chambers of the north and the chambers of the south that are in front of the Separate Place are the holy chambers where the Priests that approach the Lord shall eat the holy things. There they will lay the holy things, the meat offering and the sin offering and by the trespass offering,[550] because the place is sacred".

Verse 14: But when they have entered the Priests do not come out of the sanctuary towards the exterior atrium, but there they will leave the vestments with the ones that celebrate their ministry, because they are sacred, they will dress with other garments and thus they will go out before the people.

Verse 15: And when he had finished the measuring of the interior building he conducted me down to the way of the gate that looked towards the east and measured the structure of the House[(v')] in its perimeter for all sides.

Verse 16: He measured the side of the east with the reed and measured five hundred cubits.[(w')]

Verse 17: He encircled towards the north and measured the place of the north with a reed and measured five hundred[(x')] cubits.

Verse 18: He encircled towards the west and measured to the place of the west with a reed and he measured five hundred [cubits].

Verse 19: He encircled towards the south and measured the side of the place of the south with a reed and measured five hundred [cubits].

Verse 20: He measured the four sides of its wall, for every side in its perimeter was five hundred cubits in length and five hundred cubits[(x')] in width dividing the sanctuary and the profane place. [Towards the sanctuary there were five hundred cubits by five hundred of the square perimeter and fifty cubits of width around about the suburbs (Ezekiel 45: 2)]

Chapter 43 Verse 1: And he brought me to the gate that looked towards the east and behold the glory of the God of Israel that came from the east.

Verse 2: And his voice was like the sound of abundant water and the ground shone with its glory.

Verse 3: And I contemplated the vision which I saw, even according to the image that I had seen when it came to destroying the city and the image as the one that I had seen next to the River Chebar. And I fell face down onto the ground.

Verse 4: And the glory of the Lord entered to the House down the road of the gate that looked towards the east.

Verse 5: The Spirit elevated me and conducted me to the interior atrium, and now the glory of the Lord filled the House.

Verse 6: And I heard Him speak unto me out of the House while a man was next to me.

Verse 7: And He said to me: "Son of man, this is the place of my throne and the place of the soles of my feet where I will dwell in the midst of the children of Israel for ever and no more pollute my holy name neither of the house of Israel or his kings".

Chapter 46, Verse 19[551]: And he conducted me through to the entrance that was at the side of the gate into the sacred chambers of the Priests that looked towards the north, and there was a place in one and another side [so much to the southern side as to the northern one] that looked to the west.[(y')]

Verse 20: And he said to me: "This is the place where the Priests cook [the victim of the sacrifice] of the trespass, [by the ignorance] and of the sin offering and where they shall cook the meat offering that they do not remove anything to the exterior atrium for to sanctify the people".

Verse 21: He conducted me to the exterior atrium and he made me pass through the four corners of the atrium, and behold that there was a small atrium in each corner of the atrium.

Verse 22: In the four corners of the atrium there were small ante-rooms that were joined [of fifty cubits in square outside, but inside] of forty cubits in length and thirty in width [having removed the ten cubits in width where the steps next to the kitchen rose to the chambers of the colonnade dedicated to the banquet of the people. It agreed that those stairs were next to the kitchen so that they did not remove the food to the atria. But they would be built in the corners of the small ante-room that looked out towards the Temple so that each one of the four stairs of each one of the small ante-rooms leads to all the chambers. In the same ten cubits of the ante-room could also be built structures that served the interests of the Nazarenes, to separate the lepers and for other uses of the same way and there were steps that descended to the basements]. He measured the corners; they were the same for all the four.

Verse 23: There was a row of stones that encircled the four ante-rooms and kitchens had been built under the row.

Verse 24: And he said to me, "This is the chamber of the cooks where the servants of the House will cook the sacrifice for the people".

Ezekiel describes in addition the division of Judea into thirteen parallel parts, twelve of those, six oriented to the south and six to the north, would be given to the twelve tribes. The two parts on the boundaries of twenty-five thousand cubits in width would be given to the Principe. The part in the middle square of twenty-five thousand cubits on each side, was sacred and would be divided into three smaller parts; to the east for the Levites of ten thousand cubits in width, the west for the city of five thousand cubits and in the middle for the Sanctuary of the Priests ten thousand cubits in width, each one of them of twenty-five thousand cubits in length.

But, leaving these things, we complete the description of the Temple comparing all the Temples between themselves and supplying what Ezekiel omitted relative to the Temples of Solomon and of Herod.

One must maintain always the cubic figure (1 Kings 6: 20) of the most interior parts of the Temple since it is an image of the new city of Jerusalem (Revelation 21). Upon this and upon the holy place one must situate the chambers that together with the ground floor and the roof, would complete a height of a hundred and twenty cubits upon the pavement of the atrium (2 Chronicles 3.4). The steps of the eastern wall would rise to all the upper rooms and would be arranged with respect for a place so that the throne of God himself was not placed directly upon the ground (1 Chronicles 28: 11). As formerly in the Tabernacle, thus the lower part of the Temple should have windows; likewise lamps would be burning perpetually in the holy place, but the most interior part of the Temple was found in darkness (1 Kings 8: 2[552]; Psalms 18: 9.11 and 97: 2) except where the place had to be repaired; there the light fell from a chamber, therefore the attic windows that adorned the Temple were from the outside part. So that the absence of windows was not evident in the low part, around the side of the building was construction. It was convenient that the vestibule was lower than the remainder of the Temple so much in width as in height and did not exceed a great deal the frontal part or the architrave of wood. In front of the doorpost of this would be the columns of bronze that maintained that architrave. Of the pedestals the width may be defined as six cubits on each one of the sides, the height was twelve cubits to the highest part of the entrance; from there the height of the columns would be of eighteen cubits and that of the capitals of five cubits (1 Kings 6: 15.16) so that the total height of the base of the columns and of the capitals was thirty-five cubits (2 Chronicles 3: 15); to the architraves that added other ten or twelve cubits. On account of the height of the pedestals these columns are smaller in relation to the thinness of marble of the atria. The two pedestals, together with the intervals of their bases, were one and a half times as much[553]; they will complete the external width of the vestibule of twenty-one cubits, but with the vestibule the length of the Temple was a hundred cubits of the following form: the bases were six cubits; the front wall of the vestibule was five cubits; the remaining length of the vestibule was fifteen cubits; the front wall of the Temple was six cubits; the holy place was forty cubits; the most interior part was twenty cubits, and the western wall of the most interior part was six cubits. In total a hundred cubits,[554] to which the building of the chambers that gave to the west will add another nineteen cubits.

The former site of the temple and of the altar, the Jews, at a distance, memorised these places scrupulously so that we can remember these places; the altar would be in the middle of the Atrium of the Priests and the Temple with its chambers in the middle of the Separate Place so that there would be a hundred cubits between the curtain and the centre of the altar which extended to the vestibule towards the atrium of that and the Separate Place and between one and another, would pass the fortification of stone of a cubit in height and width and of a hundred cubits in length by the interior in each one of the seven sides. Thus we will have the form of the Sacred Precinct interior that corresponded strictly to the Tabernacle and to the atrium of this one. Solomon surrounded this Sacred Precinct with a twofold outer wall of chamber, interior and exterior, according to the number of the atria (1 Chronicles 28: 12), and he built each of the outer walls upon rows of columns (1 Kings 6: 36 and 2 Kings 11: 8 & 15), according to the description of Ezekiel. The site and the form of these colonnades was determined by the site and the forms of the gates opposite to the altar (Ezekiel 8: 3.5) because the entrances of the sides of the gates should lead directly to the promenade of the colonnades, the columns should be in front of the chambers in each of the atria and this way they would be between its axes at a distance of eleven cubits in total as it was in the Temple of Herod. For that reason the magnitude of the columns, the number and the triple row in the exterior atrium of that Temple also would be maintained here. And it is not necessary to suppose other changes in the interior atrium but the ones that are derived from the destruction of some of the gates there. In effect, in the Temple of Solomon the outer wall was built of this atrium with three series מירוש or rows of stones and with a row of cedar beams (1 Kings 6: 36). And the same thing was done in the Temple of Zerubbabel (Ezra 6: 4). Thus, it was similar to the outer wall of the interior atrium in both Temples. I suppose some false gates would have been introduced by Herod. In the books of Ezra it is said that those three rows were of cylindrical stones of form, that is to say, columns. There were two rows of columns in the colonnade under the rooms; the third was in the external facade of the exterior wall of the rooms; these corresponded with the columns of the colonnade of the Great Atrium. The series of beams of wood was found in the panelled ceiling of the colonnade, each one of them carried upon two columns and each one of these were skilfully cut so that together with the remaining one revetment of wood of the panelled ceiling presented a pleasant aspect to the ones that they looked at. Of that there was only one row of these assembled; there was only a colonnade in this outer wall. The beams did not come to a hundred that existed here in other roofs of the chambers and were made of a less rich material. Next to the external facade of the gates should be the external wall of the rooms with its half columns because, on the one hand, that facade is considered to be the end of the interior atrium (Ezekiel 40: 19), on the other hand, because the chambers were where the Priests ate the sacrifices; they were of a width of fifty cubits which corresponded to all the length of the gates. From this it is deduced that the central row of columns attached to the walls will be in front of the most exterior chambers and the most internal row will be placed central to the chambers. Between the external row and the middle are the chambers that Josephus calls "Rooms". Between the middle and the most interior rows was the colonnade. Between the most interior one and the fortification of stone was an opened-aired promenade or vestibule of the Atrium of the Priests with entrances to the sides of the gates that carried to this vestibule and to the colonnade, and with other rooms built in a double floor upon the colonnade. Thus the width of the Atrium of the Priests was the same as in the Temple of Herod, with security inside the fortifications of stone which was a hundred cubits. Inside the buildings of the rooms was one hundred and thirty-six cubits and a third, inside the walls of the colonnades was one hundred and sixty-two cubits and with the included buildings of the rooms were two hundred cubits.

Between the gate of the north and the chambers of the lower Priests which was constructed opposite the Separated Place there were thirty seven and a half cubits. In this space of the inter-columns, five (as in the Temple of Herod) will be too many and three will be too few, but four will be aesthetically pleasing and they would be of the same measurement exactly as the colonnade of the Great Atrium together with their inter-columns. Therefore it was necessary that the inter-columns would correspond with the outside wall. They will be the same measurement as the six inter-columns between the gate and the corner of the atrium towards to the east. These inter-columns will be a little greater but with an invaluable surplus that was clearly not necessary to the plan.[555] Besides, between the axes of the columns of the west and in the Great Atrium there are nine and three eighth cubits, between the east (and the axes of the columns) there was nine and seven twelfths cubits with a surplus only of one forty-fifth that divided the inter-column. There were, therefore, seven columns on both sides of the eastern gate; then also to the eastern sides of the remaining two gates in double rows towards the interior atrium, also to all the sides of those three gates in a row tos the exterior atrium. But also in the outer wall of the Great Atrium there were three times seven columns on both sides of the gate. And it is from here that Solomon, alluding to the temple that was evidently marvellous, said: "The Wisdom built her house and she had hewn out her seven columns (Proverbs 9: 1)".

From the rooms they may climb to the chambers in the corners of the atrium by the stairs that would continue to rise from the promenade in the middle of the chambers, five or six cubits in width, and illuminated by windows in the side of the stairs. There would be cubic chambers on both sides in each one of the sides which were almost nine cubits each. Thus each chamber according to the length would correspond to each one of the inter-columns, and the promenade with the walls of wood and the chambers of both sides will occupy a total width of almost twenty-four or twenty-five cubits inside the walls of the building.

For each one of the people there were two opposite chambers and below where the colonnade diminishes to the width of the chambers, a chamber would be placed upon another and would be raised by stairs up to the highest one from the lowest in a space of four or five cubits in width towards the exterior atrium. Thus a cubic chamber of the same size together with the upper stairs and the intermediate wall of wood will occupy all the width of almost fourteen cubits inside the wall of the building and in this way two chambers with their roofs will be elevated to a height of twenty-one or twenty-two cubits that was the internal height of the colonnade in the Temple of Herod. Columns of approximately seventeen or eighteen cubits would be elevated to the architrave and the architrave easily would add another four cubits more or less to the colonnade to the panelled ceiling. Thus they will be able to define symmetrically these measured. Instead of the common measurement, the width of the bases of the columns will be twenty-two palms and the interval of the base will be of a measurement with approximately half according to the length, and of two measurements according to the width; the interval of the axes of the columns will be of two measurements with approximately half according to the length and of three measurements according to the width. The height of the column will be of four measurements. Its bases will add a third part of the measurement, and the capital the third part or almost two thirds adorned with pomegranates and with the remainder of the decorations of the columns of bronze. Thus all the height of the inter-column from the pavement to the architrave is to be of five measurements, that is to say double its width between the axes of the columns. The architrave will add the sixth measurement to the beams of cedar and the internal height of the colonnade will be also double its width between the axes of the columns. Thus, as previously said, the height will be of twenty-two cubits. Besides, the floor of two cubits of thickness and the chamber as established will add another three measured of nine cubits. The ceiling with its decorations will add so much more, and the whole height of the building will be of fifteen measurements or fifty-five cubits, that is to say, triple the height of the inter-columns but equal to its length between the gate and the corner in such a way that the front facade of the chambers would be square. The structure is valued by such great simplicity and harmony of all its proportions. It descended out of the colonnade to the exterior atrium and the total height would be of almost sixty cubits, that is to say, the half of the height of the temple and equal to the height of the outer wall of the exterior atrium. The exterior and interior gates should be mutually corresponding so much in height as in length and in an equal width, and the heights of the adjacent chambers should be maintained in this way to the double-sided gates and will also be equals. The internal height of the colonnade of the one Great Atrium in the Temple of Herod will be a little over twenty-five cubits, thus far twenty-six or twenty-seven because, as in the Temple of Solomon (of whom I believe that a model had been conserved in the colonnade of Solomon to the times of Herod), the panelled ceiling of the colonnade of both of the atria might be the same height since the floor to the windows of the double-sided chambers and each one of the similar decorations of the walls of both sides they corresponded mutually so much by the equality of their heights as by the resemblance of the forms. Thus, the length of the chambers of the interior atrium, appear as if the exterior atrium, will be shown to be one and a half of the total height of these, and the length of the chambers of the Great Atrium between the gates and the corners, towards the suburbs in the direction of the Great Atrium, will be almost triple the total height.

To all the gates and corners of the Great Atrium I would add besides the height of the floor of the chambers, that is to say, three measurements or eleven cubits, so that all the height from the pavement of the one Great Atrium was almost seventy or seventy-one cubits. Therefore also this was the same as the height of the vestibule of the Temple, and the height of the gates of the one Great Atrium that was in that direction is triple the width.

The assigned heights of the construction of the chambers were also in ideal harmony with the sides of the Separate Place. These were in front of the promenade of the triple floors. The use of the promenade was, after the ascent, to go from each of the chambers. There were therefore, three floors of chambers besides the lowest floor that instead of a walkway has a space in the middle of the chambers of ten cubits in width. These chambers are equal to that of the Atrium of the Priests, and the four floors, together with the roof that occupied the space of a floor, they will arise to the height of fifty-five cubits, so that these buildings so much by their total height as by the heights of each one of the floors, will correspond exactly to the buildings of the Atrium of the Priests. But the length of a hundred cubits in front of the temple will cover in this way exactly ten chambers[556] with its walls in each one of the floors, and the length of fifty cubits of the Great Atrium will cover five chambers. These together with the side of the Atrium of the Priests will form a building of two hundred cubits of length, in whose centre the gate would be placed. And on account of the three sides of the interior of the Sacred Precinct, the southern one, the east one and the northern one, the outside of these are equal and similar, except the two atriums of the cooks which are towards the west. The promenade that was in the middle of the chambers was ten cubits in width and in a straight line with an intervening doorway of five cubits in width to the promenade of the colonnade (of the internal atrium). To both parts of the promenade a width of nineteen and a half cubits will have to be occupied by the chambers, with the exception of the width of a cubit of the fortification of stone that was joined to the chambers. That would reduce the three walkways by two cubits in width, and the exterior width of thirteen and a half cubits of the highest chambers that corresponds with the internal width of nine cubits would remain. The thickness of the front wall was two cubits in width. It adds three times two cubits and the thickness of underneath will be of almost eight cubits. In that thickness of underneath there will be storerooms constructed which was where the sacred vestments of the Priests were kept. In the external chambers towards the east there would be stairs to the walkways and in the middle looking towards the Temple and out of the central walkway the higher entrances would be opened up to the rooms of the Principum Curias.

Arranged this way, the buildings make it possible to know at the same time the distribution of the rooms between the Priests, since the classes of dignity of the rooms are in four classes. The first ones in dignity are two rooms[557] in the eastern side of the Atrium of the Priests; the second in dignity are two rooms that were to the eastern sides of the doors of the north and of the south; the third in dignity are two rooms to the western sides of these same doors. These six rooms are in the Atrium of the Priests. They follow the rooms of lower dignity to the sides of the Separate Place. And to these four dignities correspond the four dignities of the Priests. The first are the ones that precede the Guardian of the sacred things; the second is that of the Curias; the third is that of the Princeps and the fourth is of the lower Curias. Ezekiel places the Curias in the next rooms to the sides of the Separate Place. There, Ezekiel said that the Priests will eat the sacrifices and they will remove the sacred vestments. But as each Princeps of the Curias, while the Second Temple survived, was the place for the knives and the garments, thus in the first Temple they would have had their own chambers as the places that here served to keep the garments. There they had four cabinets for the garments; here they would have four chambers. This would be like that if it is thought that the shorter walkway would be for curia, two would be assigned. For the Princeps of the Curias the most appropriate places are the rooms of the third dignity; on the one hand, because the promenades of these and the promenades of the rooms of the Curias are opened between themselves and ascend to them by common stairs; on the other hand, because the number of the chambers corresponds exactly. Therefore there are twenty-four Princeps and in these two rooms twenty-four peers of chambers, four in the lower floor, four in the middle and four in the highest one to each side of the atrium; and two chambers one must assign to each one of the most important, the one that Ezekiel situates as chamber of the Principum in the exterior atrium, out of the interior gate, that is to say next to this, and the one that Jeremiah at the same time sets as the chamber of the Principe Gemariah[558] in the upper atrium next to the vestibule of the new gate in the Temple of Solomon; also the chamber of the children of Hanun,[559] next to the room of the Princeps upon the chamber of Maaseiah,[560] guardian of the gate in the same Temple, so that the room of the children of Hanum and the chamber would be in the gate that existed upon the chamber of the guardian of the gate. To the Priests of second degree of dignity one must assign the rooms of second dignity. Ezekiel distinguishes two classes of these, those responsible for the custody of the Temple and those responsible for the custody of the altar. To the first, he places them next to the gate of the north, in the room of the chambers that face the second ones, in the room of the chambers that face to the north. To the first class belonged the Overseer of the Temple (2 Chronicles 31: 13 and 35; Acts 4: 1 and 5, 24), the Overseer responsible for closing the gates, the Overseer responsible for the protection of the guards, the Overseer responsible for the loaves of bread of the proposition, the Overseer responsible for the patients, the Overseer responsible for weaving the curtains, the Overseer responsible for preparing the vestments of the priests, the Overseer responsible for repairing the buildings, etc. To the second class of overseer belonged the responsibility for describing the times, the Overseer of the singers, the Overseer responsible for the cymbals and the remaining chords,[561] the Overseer responsible for the fate, the Overseer responsible for the chickens, the Overseer responsible for the lamentation, the Overseer of the libation, the Overseer responsible for manufacturing the perfumes, and some others that are of this class together with all of these things according to Maimonides (Maimonides, *Apparatus* Temple 7). The rooms of the first class must be assigned to the most important of the first in dignity, as to the Supreme Priest and to the one that took charge of the function of the Supreme Priest when he was contaminated by the pollution or affected by the physical body, as well as when it was dedicated to the tasks of the war; these rooms are assigned also to the Vicar of the Supreme Priest, more Subvicars to whom the Great one Synedri adds with its Principe. The room of the Assistants, that is, the main headquarters of the High Priest and the chamber of first in dignity, would be placed to the right or northern side of the eastern gate, and the room of the Great Synedri to the left side. These two should be greater that the other chambers. In that, the width of the inter-column space occupied would be considered double or even triple; the width may have been enlarged also to the exterior atrium and to the roof in length and in width. The remaining gates for the Levites, for the Singers to the east, for the others other than the Supreme Priest had the most worthy part of the eastern door. And in this manner was the disposition of the interior atrium.

In the outer wall of the exterior atrium would form the chambers of a width and internal height of nine cubits and of a length four time the amount and they would be distinguished by their marble walls that would support the upper floor and the ceiling respectively. And in the middle of the chambers a continuous passage would be opened and there would be placed the tables of the people on both sides, the stairs near to the small ante-room of the cooks so that the food was not extracted to the atrium. To these responsible for the doors were assigned positions of guard to the reign of David. In the Great Atrium, next to the eastern gate, there should be situated each day six guardians of the gates under Shelemiah[562]; next to the door of the north, four guardians under Zechariah and next to the door of the south four guardians under Obed-Edom (2 Samuel 6: 11–12); likewise next to the place of the gathering two guardians, and two under the children of Obed-Edom, that is to say, two next to the room of the Assistants of the Great one Synedri, and to the west, next to the gate of Shallecheth. In the entrance of the road of the ascent that leads to the suburbs of the Temple there were six, of which four were next to the road and two next to the Parbar[563] (1 Chronicles 26: 18).

This should be understood by Parbar as the house of the lower Synedri. Therefore you should know that there were three Synedri in the Temple; the Great one Synedri with seventy-one males, and the other two with twenty-three males each. In the Second Temple situates Bartenorius to the Synedri lower next to the entrance of the Temple that is the interior of the eastern gate, after the wall of Chajil, in front of the Atrium of the Women; the Synedri medium to the right where, going for the Atrium of the Women, arrives to the entrance of the Atrium of Israel; the Synedri most supreme or Great Synedri in the room of stone. More correctly the experts in Talmud and Maimonides place the way next to the entrance of the atrium, that is to say, next to the eastern gate of the Atrium of the Women, and the lower one next to the entrance of the Mountain of the Temple, that is to say, in the east of the atrium of the Gentiles. But in the Temple of

Solomon, the Atrium of the Women was absent with its gate and the third gate was out of the Mountain of the House, next to the road of ascent; there was situated the first Synedri next to the eastern entrance of the Atrium of the Priests, the second next to the eastern entrance of the Great Atrium, the third next to the entrance of Shallecheth. And from here there were six guardians next to the eastern gate of the Great Atrium and six next to the Gate of Shallecheth. Four took charge of the protection of each one of the gates to preserve the Synedri. But the road of ascent that lead to the house of the king was so marvellous that the Queen Sheba was astounded before him (1 Kings 10: 4.5), and further down was the Gate of Shallecheth near to the house of the king. Thus, as Jehoiada[564] wanted Jehoash to be crowned king and the Temple to be defended against the attack of Athaliae, you will be able to determine that a third part of the town would be next to the gate of the soldiers or the Entrance, a third part next to the gate (מוּד or more correctly סוּר) of the separation and a third part next to the house of the King (2 Kings 11: 2; 2 Chronicles 23).

Where the guardians of the gates usually placed themselves, there the people were placed in such a way that when the guards of the Temple closed the gate, according to

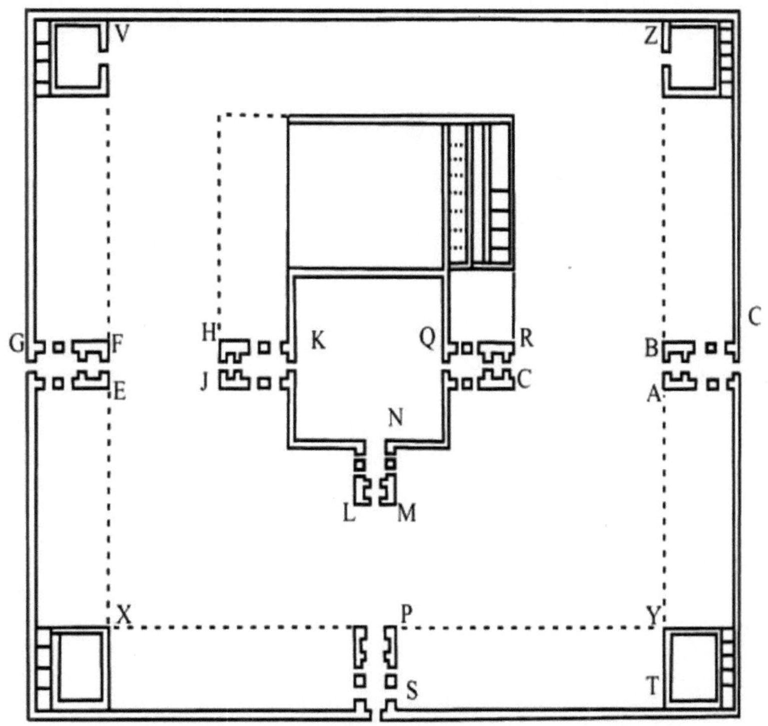

Fig. 5 Copy of a ground plan of the Temple of Solomon included in the manuscript. The letters on the plan have no correlation with the text[565] (Drawn by the author from Isaac Newton, *Introduction to the Lexicon of the Prophets, Part Two: About the Appearance of the Jewish Temple* (Babson Ms 0434) (unpublished manuscript, Babson College) fol. 70r.)

custom, the eastern gate, two parts of the people were situated next to the other two gates of the Great Atrium guarding the Temple (2 Kings 11: 7) and the third part, next to the door of Shallecheth next to the house of the king, guarded by the suburbs. Of where it is deduced clearly that there was only a gate of the suburbs, that this was situated to the west and that the western side of the great one atrium lacked a gate. Therefore, on having gone to the house of the king, not all of the people were going out of the gate of the soldiers (2 Chronicles 23: 20) (Fig. 5).

Notes

(a) In the Hebrew text already divulged is found כנגמ "to the south" but in *Septuaginta* "north" and the ancient texts read דגנמ "of the opposite part".

(b) The tradition of the Jews is unanimous that the current cubit was formed by five palms and the sacred one of six. Thus the Angel uses the reed of six sacred cubits. Jerome was reading a corrupted version which he took to serve as an example whereby the use of the reed was of six cubits and a palm, and the Latin ones followed him ordinarily. In the Hebrew text it is written that the reed was of six cubits and a palm and thus also reads in Jonathan. In this manner without a doubt he was reading *Septuaginta,* although formerly in them it was written with the sense changed παλαισής instead of παλαισή. But also Ezekiel in another place, says explicitly that its cubits were comprised of a cubit and a palm (43: 13), that the reed was of six cubits exactly (41: 8) and that the chambers of a reed (40: 7) were of six cubits exactly (verse 12). And the whole width of the gate that is formed with two reeds of the two chambers and with the intermediate width of thirteen cubits was of twenty-five cubits and all the length of the gate that was formed with the three reeds of the three chambers, with the two reeds of the two colonnades with the five cubits twice between the chambers, say ten cubits of the doorpost, of this went in total of fifty cubits.

(c) In the translation of *Septuaginta* the number seven is read and appears opposite in the following (22: 26).

(d) In the Hebrew text already is added "And a threshold is one reed in width"; these words are lacking in the copy of the *Septuaginta Virorum* and are the repetition of the immediately preceding words with the same number of syllables, except that instead of השעד "gate" by mistake subsequently was written אחד "one".

(e) Thus reads the *Septuaginta*. And afterwards you consider thirty is spoken of these vestibules as if they had been mentioned previously. Further unoccupied are the intermediate spaces on the way to the colonnades of the chambers **EVTF, GSRH** and it is clear because the Angel measured the width of the chambers **EV, FT** and the width of the gate **Vv**, on the other hand, because the Priests passed from these intermediate spaces of the gates of the interior atrium from their storeroom towards the interior atrium, they returned from there towards their storerooms and meanwhile they did not pass to the exterior atrium (42: 14) and besides Ezekiel entered and advanced towards the northern gate of the chambers of the Priests (46: 19).

(f) Here in the Hebrew text the scribe by mistake twice wrote the words, "inside with a reed and measured the vestibule of the gate," because he had the same syllables. In the copies of Jerome and in the authors of the Syrian version were read only one time. In the version of *Septuaginta* up to the Roman edition they are omitted completely by the contrary error of a scribe, but in the Alexandrian Codex and in the Arabian version they are read once.

(g) Here he describes the place of the vestibule and of the chambers as the opposite to that described in verse 31; later he described the gates of the interior atrium.

(h) This width was formerly in the translation of *Septuaginta*, as was testified by Jerome. Nevertheless he thought that it was the reading that he considered to be less correct. Therefore, he says that the Scriptures do not have to say in the same place that the width was of ten and of thirteen cubits. Others thinking the same thing they corrected recklessly the place so that instead of "width" it reads "length"; subsequently in the version of *Septuaginta* as in the Hebrew text. But to be read in the same manner convinces me that one must read "width" from the internal width of the gate to be thirteen cubits (as turns out to be evident upon reducing the width of the two chambers from the total width **Vv** of twenty-five cubits) and the measurement of its internal width itself is not expressed in any other part.

(i) Where now reads ששים "sixty," *Septuaginta* correctly reads (twenty). Jonathan and the Syrian version understand this to be the height of the front or of the doorposts of the gate. And surely the height should be expressed in some part, but it is not expressed elsewhere. But the width of the doorway was of ten cubits and the height according to the rules of the architects, should be double the width. Thus also the gates of this Temple will be square in magnitude with the gates of the second Temple.

(j) In the Hebrew text: "gate next to the doorpost of the atrium," is singular in number instead of being plural. Thus, next to the doorpost (verse 16) and next to the wall side (verse 18) instead of next to the doorposts and next to the wall sides. But surely it is here in front of the doorway of the gates that stand open **VT SR, Pp**, in the same way as in the following verse.

(k) דצף signifies the pavement was covered with stones (2 Chronicles. 7: 3) and glowing coal (Isaiah 6: 6 and 1 Kings. 19: 6). And from there they came upon the sight of an oven, the place where it is found was the open space of the oven. Thus in *The Songs of Solomon* 3: 10 the description of the bed of Solomon spoke of its columns of silver, its head of gold, being covered with purple and the central part רצוף אהבה "a covered space between the columns or gate of the love". Thus furthermore in Ester 1: 5.6 the real atrium is described as being adorned with columns of marble, upon paving with columns where the beds for the banquets were found; it is called רצפה. And thus in this passage of Ezekiel *Septuaginta* translates רצפה περιστυλα what the Latin translator, according to Jerome translates "between the columns" while we translate it as "pavement between the columns," the support for the columns or the place of the covered columns. And this interpretation confirms, being the same as, Ezekiel (42: 6) upon speaking of the columns of the atria.

(l) Thus they were reading the *Septuaginta*. In the Hebrew text it appears as so great and "toward the north". Villalpando upon referring these words to the preceding phrase, imagined that the Angel measured it so that the length of the atrium from the south to the north was in width between the gates (a hundred cubits and another) and for that reason he incredibly thought that all the atrium was divided into nine small ante-rooms, each one of them of a hundred cubits long and of the building fifty cubits wide. But in the verses 23 and 27, where are repeated measurements that are similar, it is evident that the Angel only measured the distance between the gates and that those words "and toward the north" belonged to the following phrase. Therefore the Prophet says that when he was lead out of the eastern gate towards the exterior atrium, he saw thirty chambers upon the pavement with columns in the surroundings of the atrium. If by exterior atrium you understand here, as Villalpando did, to be a small atrium, they would have to have enumerated seven exterior ante-rooms; in spite of that the Angel did not speak about them in any other place. And besides, around the small atrium there should be thirty chambers, which are the same ones that Villalpando recognizes are impossible if the spaces of the gates are not to be counted; against this are the express words of the Prophet that place the pavements with their thirty chambers next to the wall sides of the gates and in front of the length of the gates. Because, by exterior atrium it is not to be understood here as a small atrium but a large atrium, and if we situated it around the pavement of columns with its chambers (therefore thus they were situated in the second Temple), this would be to remove all the obstacles of the small ante-rooms so that the Prophet could see and enumerated the chambers when he was lead from the gate towards the atrium. But why argue against an opinion that itself is not founded but for the words "and toward the north" but is an incorrect translation? If the small ante-rooms had been so many then the Angel would have spoken of them in other places? Would not he have done it with clearer words? Certainly we cannot interpret those words as Villalpando does, unless we can expand its sense falsely with some words in this way; and [the width of the atrium to the south] to the north. And why then do we not expand likewise thus; and [conducted me] "toward the north," especially because he thus reads the old *Septuaginta.* And although the words "conducted me" were already absent in the Hebrew text, however, they can and should be supplied from other analogous places (compare verses 24, 27, 32, 35, 48, etc.). Therefore the Angel here passes with the Prophet from one gate to another and how often has he made the passage, "he leads me". It happens equally if you think tenaciously that one must follow the Hebrew text of today, therefore this text without the words leads me to this translation thus: "And toward the north was also the gate that looked towards the direction of the north in the exterior atrium," etc., or omitting the second as redundant: and "toward the north was the gate that", etc. To such an extent does it not agree that if we follow the statement of Villalpando it has no support and it is lacking in reason. And likewise I cannot follow it unless we want to move away from the proportions of Moses in the atrium that surrounds the immediate temple and the altar which was established by Villalpando himself as being a length over the double one of the width.

(m) ו instead of the *Septuaginta* reads here as כ.

(n) Thus reads *Septuaginta* and this way reads verses 21, 29, 33, 36.

(o) The verse 30 is missing in *Septuaginta* in the Roman edition; however, in the Aldine edition, in the Alexandrian manuscript and in the Arabian version they followed the Hebrew text with its Syrian and Chaldaean versions.

(p) *Septuaginta* reads "gate" instead of more correctly "atrium," therefore the Prophet already had been lead into the atrium before.

(q) Thus *Septuaginta* and the Latin translation; also in the analogous places (verses 22, 25, 29, 33) that is confirmed in the reading.

(r) The Hebrew text now reads ואילו "and its doorposts". But *Septuaginta* and Jerome were formerly read correctly as ואילמו "its vestibule," as clears up in the analogous places (verses 31 and 34 and not, 9, 22 and 26).

(s) Thus interprets Jonathan in the *Targum* the word שפתים.

(t) לשכה signifies that there were many "chambers" in the "room". Here and in the following places, you should understand "the building of many chambers," to be read in the same way as Josephus who calls the "room" the whole construction between the two gates.

(u) שרים sometimes the "singers" often signifies the presence of the "Principals". To the singers themselves would not be assigned such noble rooms in the Atrium of the Priests. With the Syrian translation, because of the Principals in the Temple of Solomon, there was a room of the Principals (Jeremiah 35: 4) and the room or chamber of the Principal Gemariah in the upper atrium (Jeremiah 36: 10 & 12).

(v) In the Hebrew text it is written אשר "anyone," but in the *Septuaginta* it reads אחר "one"; they are in agreement in the sense of it.

(w) In the Hebrew text it reads הקדים "of the east," but *Septuaginta* read הדרים "of the south" and then the same translations; also the opposition between the gates of the south and of the north, about which is the place he is conducted to, is confirmed in the reading.

(x) *Septuaginta* reads "twelve". The situation of the vestibule can be discussed, perhaps it is necessary to place the length of this according to the width of the Temple, as it seems at first sight that this is affirmed in 1 Kings 6: 3. But the facades of the buildings should be wider than the vestibules, and the fact that the gate of the vestibule should be less than that of the Temple confirms that the former wall of the vestibule was narrower than that of the Temple, and in the same way that the vestibule, in the place described to us, responds better to the images of the three colonnades on the three sides. And, together with the columns of bronze, complete the length of the temple of a hundred cubits, as agrees.

(y) In the Vulgate translation for אשר "that", *Septuaginta* reads רשע "ten," coinciding with Theodotion in this regard. Theodotion and north close to the second edition, Jerome said that they read רשצ תשצ ' "eleven".

(z) In Vulgate תאהל, "the tent," in *Septuaginta* האיל "the doorposts": this is what the sense demands.

(a′) Vulgate ורחב "and the width". *Septuaginta* interprets וכתפ with a better sense and with measurement analogous and in accordance with verse 2.

(b′) Thus read *Septuaginta* in agreement with the sense.

(c′) In the Hebrew text "were thirty-three" one over another (therefore that word is punctuated מי מעפ]. And as for the *Septuaginta* it translates τριάκοντα τρìς δίς, "thirty, three times doubled" for one over another, that is to say, thirty according to the length, three times according to the height, twice according to the width. But this translation; however, interprets the order of words badly, therefore τρίς or (as they have in some copies) τρία belongs to the preceding words "the rooms were above the rooms" and signifies triple floor, as is clear also in verse 16.

(d′) In the Hebrew text "what was coming" that is to say down from the wall jutting out and coming nearer. With the contrary expression but with the same sense, in Latin it would be the receding or decrease of the walls.

(e′) In the Hebrew text it states "highest".

(f′) In the Hebrew text it already lacks ו and they where reading *Septuaginta*, the Latin and Jonathan version.

(g′) ב׳ח עלשה אשד לב׳ח ובין חלשכוח instead of ב׳ח reads כ׳ן with *Septuaginta* it translates to ἀναμέσον τών πλερών τού οίκου καί άναμέσον τών έζεδρών "between the room next to the House and between the rooms that is to say, between the sides and the chambers". The Hebrew expresses the word "between" two times, we only express it once. Again one must note here the distinction between the chambers situated to either side of the remaining space. To some they are called צלעוח "sides or side chambers," to others לשכוח "rooms" or "simply chambers". It is said that these are next to the House; they are not. This is so that clearly you understand that the temple was surrounded by two types of chambers in each one of the floors, one closer, the other was further away, in the middle there was the promenade or corridor, which is called the remaining space. The diverse names and the different places show evidently that the floors of the chambers were diverse.

(h′) The word "et" here reads in the *Septuaginta* according to the Alexandrian manuscript and the Arabian version. Again the width of twenty cubits was not the width of the remaining space or of the space that was between the chambers (as some have imagined), therefore this space is said to be of five cubits in width, but here the width was that of the Separate Place that separates the side building and the chambers of the Priests where the sacrifices are consumed.

Thus, one has not yet spoken of these chambers; the description of these begins after Chapter 42 and, besides, if a hundred cubits in width of the atrium that was between these chambers seventy cubits of the width of the Temple were removed and of the side of the building jointly (Ezekiel. 41: 12, 14, & 15), there would remain thirty cubits, fifteen on one side and fifteen on the other, as the width of the Separate Place, that consequently is not of twenty cubits. But, why refer then to those twenty cubits? I will tell you. The Prophet writes that those chambers of the Priests were situated facing the twenty cubits of the atrium interior and in front, upon the pavement of columns of the exterior atrium (42: 2, 3), as the pavement of columns describes a part of the colonnade opposite of the exterior atrium, thus the twenty cubits, on the other hand, describes the opposite building of the interior atrium and are therefore according to its width. This is also evident upon doing the computation. Therefore if all the previously said widths of seventy cubits is subtracted, the width of the Temple of twenty cubits and the width of their wall of five cubits of the side and five cubits of the other, to the first reduction, there will remain forty cubits, twenty on one side and twenty on the other as the width of the perimeter of the adjacent building to the Temple. Or thus the width of the side chamber is five cubits, as above it. That of the remaining space is of five cubits (Ezekiel, 41: 11). That of the exterior chambers as that of the interior and that of the external wall is also of five cubits (Ezekiel 41: 9.12), that is to say, in total twenty cubits, as I said above. I have explained all these things extensively to leave out the errors of some versions that I do not stop in counting. Finally, with the words "twenty cubits" the same building was appointed while the Temple of Solomon survived. Solomon built it to be twenty cubits [that is to say, a building of twenty cubits] to the side of the House; he covered it with boards of cedar from the floors to the ceilings (1 Kings 6: 16).

(i′) קיתא signifies "the gallery removed to the building and encloses him" and indicates the gallery that was around the Temple of fifteen cubits of width extending to the one that is called the

Separate Place and in the direction following the promenade between the chambers built in three floors around the Temple and in Chapter 42: 3 & 5; down from the promenade these were in the inlets of the walls next to each one of the floors of the chambers of the Priests.

(j′) Thus it is said in *Septuaginta.*

(k′) In the work of Jerome it read ריקב "in the wall" and לכיחה "the Temple" only on one occasion did this come to his mind.

(l′) In the Hebrew text: וכרא "its length, " in the *Septuaginta* ונרא its "base" or "stylobate" was in harmony with the sense.

(m′) Thus were also the gates of the Temple according to those described by the experts in the Talmud.

(n′) Also there were the beams in the vestibule of the second Temple but Ezekiel lacks any account of their arrangement in the form of the colonnade of columns whose description can be seen in the experts in the Talmud.

(o′) Thus *Septuaginta* according to the Alexandrian manuscript and the Arabian version, reads this with the Hebrew text vulgate; you would find הכשלה "room," is understood for the room of the whole building of the chambers of a hundred cubits in length.

(p′) In Vulgate תחא המא דרך "space of a cubit". In the *Septuaginta* חמא חאמ דרא "of a length of a hundred cubits".

(q′) In Vulgate החצד "of the atrium". In the *Septuaginta* דלהמ "of the promenade".

(r′) Thus it is said in *Septuaginta.*

(s′) In Vulgate דכחר ןכ "thus was the width of these". In the *Septuaginta.* וכרחכן "and according to the width of these".

(t′) In Vulgate וכל "and all". In the *Septuaginta* וככל "and according to all".

(u′) In Vulgate יחחפכו "and thus the doors". In Jerome's and the Syrian version יחחפכ "thus the doors". In the *Septuaginta* it is lacking.

(v′) In the Hebrew text it lacks this. In the *Septuaginta* it has ὓπόδειγμα τού οἵκου "model" or "structure" of the building.

(w′) In vulgate םינק חומא שמח "five cubits of reeds," altering the sense. In the *Septuaginta* תואמ שמח "five hundred" and according to the Arabian version תומא תואמ שמח "five hundred cubits," that is the best reading. Further, the preceding words "with a reed to measure" which are read in the Hebrew text are redundant and they are not read in *Septuaginta*.

(x′) In the Hebrew text it read "reeds" while in the *Septuaginta* it read "cubits". But in *Septuaginta* they are cubits but it should not be read as reeds, this is explained extensively by L. Cappel (in Walton in the Prolegomena of the *Biblia Polyglotta)* who was in dispute against the opinion of the fantasy of Villalpando, and the cubits are computed through parts thus: the length of the exterior gate situated towards the north was fifty cubits; from there to the interior gate was a hundred cubits; the length of the interior gate was fifty cubits; the width of the interior atrium was a hundred cubits; the length of the southern interior gate was fifty cubits; and from there to the southern gate outside was a hundred cubits the length of the gate. In total it was five hundred cubits. This is the width of the atrium that was between the external facades of the external gates and, therefore, was of five hundred cubits the exterior length by each one of the sides of the external wall, whose height and width were a reed. And, thus, the perimeter of this wall was the one that the Angel measured. He added fifty cubits for all parts towards the suburbs, and each side of the square was of six hundred cubits.

(y′) The order of these periods in the Hebrew text vulgate is changed. But the Angel did not cross over through the narrow passage from the side of the north next to the south, as is said there, but measured the border in succession of the four sides, according to the old text of *Septuaginta.*

Notes

Introduction

1. Michael White, *Isaac Newton: The Last Sorcerer* (London: Fourth Estate, 1998), p. 1.
2. David Clark and Stephen P. H. Clark, *Newton's Tyranny* (New York, NY: W. H. Freeman and Company, 2000), p. viii.
3. Albert Einstein, "Introduction", in *Opticks*, ed. Albert Einstein (New York, NY: Dover Publication, 1952), pp. lix–lx.
4. see Patricia Fara, *Newton: The Making of Genius* (London: Picador, 2002).
5. William Stukeley, *Memoirs of Sir Isaac Newton's Life* (London: Taylor and Francis, 1936).
6. David Brewster, *Memoirs of the Life, Writing and Discoveries of Isaac Newton* (Edinburgh: Thomas Constable & Co, 1855).
7. William Stukeley, *Memoirs of Sir Isaac Newton*, p. 20.
8. N. W. Chittenden, "The Life of Sir Isaac Newton", in *Newton's Principia: The Mathematical Principles of Natural Philosophy*, eds. Andrew Motte and N. W. Chittenden (New York, NY: Daniel Adee, 1848), p. 58.
9. Brewster, *Memoirs of the Life, Writing and Discoveries of Isaac Newton,* p. 138.
10. Anonymous, "General News", *Plymouth and Cornish Advertiser* 1870, in *Isaac Newton: Eighteenth Century Perspectives*, ed. A. Rupert Hall (Oxford: Oxford University Press, 1999), p. 175.
11. Brewster, *Memoirs of the Life, Writing and Discoveries of Isaac Newton*, p. 138; Gale E. Christianson, *Isaac Newton* (Oxford: Oxford University Press, 2005), p. 85.
12. Robert M. Hazen, *The Diamond Makers* (Cambridge: Cambridge University Press, 1999), p. 9.
13. Fara, *Newton: The Making of Genius*, Fig. 7.6.
14. James Hamilton, *London Lights: The Minds that Moved the City that Shook the World* (London: John Murray, 2008) p. 174.
15. Derek Gjertsen, *The Newton Handbook* (London & New York: Routledge & Kegan Paul, 1986), p. 426.
16. Frank E. Manuel, *The Religion of Isaac Newton* (Oxford: The Clarendon Press, 1974), p. 14.
17. Anonymous, "Art and Literary Gossip", *Manchester Times,* 1876.
18. Jose Faur, "Newton, Maimonidean", *Review of Rabbinic Judaism* 6, no. 2/3 (2003), p. 228.
19. Ibid, p. 229.
20. John Maynard Keynes, "Newton the Man", in *Essays in Biography*, ed. John Maynard Keynes (Cambridge: Cambridge University Press, 1972), p. 363.
21. Ibid, pp. 363–364.
22. Anonymous, "Sir Isaac Newton Tercentenary", *The Times* 1946, article 336, col B.
23. Keynes, "Newton the Man", p. 363, note 1.
24. Ibid, p. 366.
25. Ibid, p. 374.

26. Gjertsen, *The Newton Handbook*, pp. 88–90; Frank E. Manuel, *A Portrait of Isaac Newton* (Cambridge, MA: The Belknap Press of Harvard University Press, 1968), pp. 213–225; White, *Isaac Newton: The Last Sorcerer*, pp. 222–253.
27. Newton as quoted by Stephen D. Snobelen, "To Discourse of God: Isaac Newton's Heterodox Theology and His Natural Philosophy", in *Science and Dissent in England, 1688–1945*, ed. Paul Wood (London: Ashgate, 2004), p. 39; Jose Faur, "Newton, Maimonides, and Esoteric Knowledge", *Cross Currents* 40, no. 4 (1990), p. 546.
28. Isaac Newton, "Optics", in *Mathematical Principles of Natural Philosophy*, ed. Florian Cajori (Chicago, IL: The University of Chicago, 1952), p. 370 (Newton's italics).
29. Richard H. Popkin, "Newton's Biblical Theology and His Theological Physics", in *Newton's Scientific and Philosophical Legacy*, eds. P. B. Scheurer and G. Debrock (Dordrecht: Kluwer, 1988), p. 93; Richard H. Popkin, "Some Further Comments on Newton and Maimonides", in *Essays on the Context, Nature, and Influence of Isaac Newton's Theology*, eds. James E. Force and Richard H. Popkin (Dordrecht: Kluwer, 1990).
30. Manuel, *A Portrait of Isaac Newton*, p. 373; Thomas C. Pfizenmaier, "Was Isaac Newton an Arian", *The Journal of the History of Ideas* 58, no. 1 (1997); Gjertsen, *The Newton Handbook*, p. 590.
31. Markley, "Newton, Corruption, and the Tradition of the Universal History", in *Newton and Religion: Context, Nature, and Influence*, eds. James E. Force and Richard H. Popkin (Dordrecht: Kluwer, 1999), pp. 122–128.
32. An anti-Trinitarian sect that arose in the Reformation and was popular in the late seventeenth century.
33. Keynes, "Newton the Man", p. 368.
34. Popkin, "Newton's Biblical Theology and His Theological Physics", p. 84.
35. Matt Goldish, "Newton on Kabbalah", in *The Books of Nature and Scripture: Recent Essays on Natural Philosophy, Theology, and Biblical Criticism in the Netherlands of Spinoza's Time and the British Isles of Newton's Time*, eds. James E. Force and Richard H. Popkin (Dordrecht: Kluwer, 1994), p. 92.
36. James E. Force, "Newton, the Lord God of Israel and Knowledge of Nature", in *Jewish Christians and Christian Jews*, eds. Richard H. Popkin and G. M. Weiner (Dordrecht: Kluwer, 1994), p. 133.
37. Faur, "Newton, Maimonides, and Esoteric Knowledge"; Faur, "Newton, Maimonidean"; Popkin, "Some Further Comments on Newton and Maimonides".
38. Faur, "Newton, Maimonides, and Esoteric Knowledge", p. 526.
39. Maimonides as quoted by Ibid, p. 526.
40. See Chap. 4.
41. John R. Harrison, *The Library of Isaac Newton* (New York, NY: The Cambridge University Press, 1978), pp. 74–75.
42. Manuel, *The Religion of Isaac Newton*, p. 84; Matt Goldish, *Judaism in the Theology of Sir Isaac Newton* (London: Kluwer , 1998), p. 18; Richard S. Westfall, *Never at Rest: A Biography of Isaac Newton* (Cambridge: Cambridge University Press, 1980), p. 346; Faur, "Newton, Maimonides, and Esoteric Knowledge", p. 526.
43. Harrison, *The Library of Isaac Newton,* Catalogue number 216.
44. For a full discussion on these books see Goldish, *Judaism in the Theology of Sir Isaac Newton,* pp. 26–32.
45. Harrison, *The Library of Isaac Newton*, Catalogue numbers 1018–1022.
46. Ibid, p. 258.
47. Isaac Newton, "Yahuda Ms 1.1", (unpublished manuscript, Jerusalem: Jewish National and University Library, undated), p. 14r.
48. A. Rupert Hall, *Isaac Newton: Adventurer in Thought* (Cambridge: Cambridge University Press, 1992), p. 13; Westfall, *Never at Rest: A Biography of Isaac Newton*, pp. 82–83; Maurizio Mamiani, "The Rhetoric of Certainty: Newton's Method in Science and in the Interpretation of the Apocalypse", in *Persuading Science: The Art of Scientific Rhetoric*, eds. Marcello Pera and William R. Shea (USA: Science History Publications, 1991), p. 168.
49. Mamiani, "The Rhetoric of Certainty: Newton's Method in Science and in the Interpretation of the Apocalypse", p. 162.

50. Ibid; The sixteen rules are outlined in Newton, "Yahuda" Ms 1.1, pp. 12r–20r. These rules will be expanded upon in Chap. 3.
51. Mamiani, "The Rhetoric of Certainty: Newton's Method in Science and in the Interpretation of the Apocalypse", pp. 163–164.
52. Newton as quoted by Ibid, p. 168 originally from University Library, Cambridge MS Add. 3968, p. 101.
53. Ibid, pp. 168–169.
54. Newton as quoted by Force, "Newton, the Lord God of Israel and Knowledge of Nature", pp. 147–148.
55. Stuart Ross Taylor, *Solar System Evolution* (Cambridge: Cambridge University Press, 1992), p. 8.
56. David Kubrin, "Newton and the Cyclical Cosmos: Providence and the Mechanical Philosophy", *Journal of the History of Ideas* 28, no. 3 (1967), p. 332.
57. Snobelen, "To Discourse of God: Isaac Newton's Heterodox Theology and His Natural Philosophy", p. 54.
58. Andrew Cunningham, "How the Principia Got Its Name: Or Taking Natural Philosophy Seriously", *History of Science* 29 (1991).
59. Ibid, p. 388.
60. Newton as quoted by Richard S. Westfall, "Isaac Newton's Theologiae Gentilis Origines Philosophicae", in *The Secular Mind*, ed. W. Warren Wagar (New York and London: Holmes & Meier 1982), p. 16.
61. Isaac Newton, *De Gravitatione et Aequipondio Fluidorum* (unpublished manuscript, Cambridge: Cambridge University Library).
62. Betty Jo Teeter Dobbs, *The Janus Faces of Genius* (Cambridge: Cambridge University Press, 2002), p. 139.
63. A. Rupert Hall and Marie Boas Hall, eds., *Unpublished Papers of Isaac Newton* (Cambridge: Cambridge Press, 1962), p. 137.
64. Ibid, p. 139.
65. Scott Mandelbrote, " 'A Duty of the Greatest Moment' Isaac Newton and the Writing of Biblical Criticism", *The British Journal for the History of Science* 26, no. 3 (1993), p. 283.
66. Isaac Newton, *The Original of Religions Yahuda Ms 41* (unpublished manuscript, Jerusalem: Jewish National and University Library, undated), p. 7r.

Chronology, *Prisca Sapientia* and the Temple

67. Goldish, *Judaism in the Theology of Sir Isaac Newton*, p. 26.
68. Harrison, *The Library of Isaac Newton*.
69. Ibid, p. 1; Mordechai Feingold, "Catalogue of the Libraries of Jerome and John Reynolds", in *Private Libraries in Renaissance England*, eds. R. J. Fehrenbach and E. S. Leedhan-Green (Binghamton, NY: Medieval & Renaissance Texts & Studies, 1993), pp. 85–111; Claire Tomalin, *Samuel Pepys: The Unequalled Self* (London: Penguin, 2002), plates 43, 44.
70. Ibid, p. 19.
71. Stukeley, *Memoirs of Sir Isaac Newton*, p. 70. Stukeley's manuscript was dated 1752, however, it was not printed until 1936.
72. Goldish, *Judaism in the Theology of Sir Isaac Newton*, p. 26.
73. Ibid.
74. Richard S. Westfall, *The Life of Isaac Newton* (Cambridge: Cambridge University Press, 1993), p. 24; Isaac Newton, *Notebook Containing Newton's Expense Accounts from 19 March* 1659 *through His Cambridge Years* (unpublished manuscript, Cambridge: Trinity College Library, c1659), fol. v.
75. Harrison, *The Library of Isaac Newton*.
76. H. W. Turnbull, ed., *The Correspondence of Isaac Newton, Volume 2, 1676–1687* (Cambridge: Cambridge University Press, 1960), pp. 297–298.

77. Ibid, p. 300.
78. Frank E. Manuel, *Isaac Newton: Historian* (Cambridge: Cambridge University Press, 1963), p. 2.
79. Vaughan Hart, *Art and Magic in the Court of the Stuarts* (London: Routledge, 1994), p. 32.
80. Kenneth J. Knoespel, "Newtonian the School of Time: The Chronology of the Ancient Kingdoms Amended and the Crisis of the Seventeenth Century Historiography", *The Eighteenth Century* 30, no. 3 (1989), p. 26.
81. George Sarton, *Ancient Science through the Golden Age of Greece* (London: Courier Dover, 1993), p. 143.
82. Manuel, *Isaac Newton: Historian*, p. 2.
83. Westfall, *Never at Rest: A Biography of Isaac Newton*, p. 805.
84. Ibid.
85. Manuel, *Isaac Newton: Historian*, p. 22.
86. Ibid, p. 23.
87. H. W. Turnbull, ed., *The Correspondence of Isaac Newton, Volume Five, 1694–1709* (Cambridge: Cambridge University Press, 1960), pp. 330–331.
88. Manuel, *Isaac Newton: Historian*, p. 809.
89. C. W. Ceram, *Gods, Graves and Scholars: The Story of Archaeology* (London: Sidgwick and Jackson Limited, 1952), p. 4.
90. Ibid, p. 85.
91. J. Arthur R. Munro, "Notes on the Text of the Parian Marble II", *The Classical Review* 15, no. 4, pp. 355–361 (1901).
92. Martin Gorst, *Measuring Eternity: The Search for the Beginning of Time* (New York, NY: Broadway Books, 2001), pp. 26–28.
93. Isaac Newton, "A Short Chronicle from the First Memory to the Conquest of Persia by Alexander the Great", in *The Chronology of Ancient Kingdoms Amended* (London: Histories & Mysteries of Man, 1988), p. 19.
94. William Whiston, "An Abstract of the Consultation of Sir Isaac Newton", *Present State of the Republick of Letters* 3, April (1729), p. 247.
95. Anonymous (W. X), "A Critical and Apologetical Dissertation for Sir Isaac Newton's New System of Chronology and Mythology; Especially, as Compared with That of the Late Bishop Cumberland, & C", *Present State of the Republick of Letters* 2, November (1728); Anonymous (W. X), "A Critical and Apologetical Dissertation for Sir Isaac Newton's New System of Chronology and Mythology; Design'd Chiefly as an Introduction to a Farther Enquiry and Examination into Its Truth and Certainty", *Present State of the Republick of Letters* 2, September (1728).
96. Isaac Newton, *The Chronology of Ancient Kingdoms Amended* (London: Histories and Mysteries of Man, 1988), pp. 94 and 100–101.
97. Ibid, p. 66.
98. Ibid, p. 191.
99. A. Rupert Hall and Laura Tilling, eds., *The Correspondence of Isaac Newton, 1718–1727* (London, New York, Melbourne Cambridge University Press, 1977), p. 279.
100. Manuel, *Isaac Newton: Historian*, p. 25.
101. Westfall, *Never at Rest: A Biography of Isaac Newton*, p. 809.
102. Isaac Newton, Remarks on the Observations made on a Chronological Index of Sir Isaac Newton *Philosophical Transactions of the Royal Society* 33, 1725, pp. 315–321.
103. Ibid, p. 316.
104. Clark and Clark, *Newton's Tyranny.*
105. Jason Socrates Bardi, *The Calculus Wars: Newton, Leibniz, and the Great Mathematical Clash of All Time* (New York, NY: Thunder's Mouth Press, 2006).
106. James Ussher, *The Annals of the World* (London: Printed by E. Tyler, for J. Crook and for G. Bedell 1658).
107. Gorst, *Measuring Eternity: The Search for the Beginning of Time,* p. 11.
108. Ibid, p. 42.

109. Stukeley, *Memoirs of Sir Isaac Newton*, p. 62.
110. William Stukeley, *The Family Memoirs of the Rev William Stukeley MD and the Antiquarians and Other Correspondence of William Stukeley, Roger and Samuel Gales Etc* (Edinburgh: William Blackwood, 1883), p. 262.
111. Souciet as quoted by Manuel, *Isaac Newton: Historian*, p. 33.
112. Anonymous, "Essays on Chronology; Being a Vindication of the System of Sir Isaac Newton, by a Member of the University, Cambridge", *Gentleman's Magazine* December (1827), p. 520; Mr. Urban, "Remarks on Sir Isaac Newton's Chronology", *Gentleman's Magazine* June (1826), pp. 593–595.
113. Westfall, *Never at Rest: A Biography of Isaac Newton*, p. 873.
114. Gjertsen, *The Newton Handbook,* p. 399.
115. Ibid, pp. 88–90; Manuel, *A Portrait of Isaac Newton*, pp. 213–225; White, *Isaac Newton: The Last Sorcerer,* pp. 222–253.
116. White, *Isaac Newton: The Last Sorcerer*, p. 262.
117. John Conduitt, "Memoirs of Sir Isaac Newton", in *Collections for the History of the Town and Soke of Grantham Containing Authentic Memoirs of Sir Isaac Newton*, ed. Edmund Turner (London: William Miller, 1806), p. 162; John Arbuthnot' s *Tables of Grecian, Roman, and Jewish measures, weights and coins; reduced to the English standard* was published in 1705, 1707, 1709, 1727 and 1747.
118. Clark and Clark, *Newton's Tyranny*, p. 116.
119. David Castillejo, *The Expanding Force in Newton's Cosmos* (Madrid: Ediciones De Arte Y Bibliofilia, 1981), p. 82.
120. Newton, *The Chronology of Ancient Kingdoms Amended,* p. 28.
121. Ibid.
122. Manuel, *Isaac Newton: Historian*, pp. 48–49.
123. Bishop of Rochester, "An Account of Want Related to the Publishing of Sir Isaac Newton's Chronology of Ancient Kingdoms", in *The Chronology of Ancient Kingdoms Amended* (London: 1770), pp. 7–8.
124. Ibid, p. 7.
125. Newton, *The Chronology of Ancient Kingdoms Amended*, p. 43.
126. Ibid, p. 46.
127. Ibid, p. 52.
128. Ibid, p. 127.
129. Ibid, pp. 94 and 100–101.
130. Ibid, p. 71.
131. Ibid, p. 71.
132. Ibid, p. 63.
133. Ibid, p. 81.
134. Ibid, p. 57.
135. Ibid, p. 126.
136. Ibid, p. 67.
137. Ibid, p. 191.
138. Newton as quoted by Manuel, *Isaac Newton: Historian,* note 6, p. 269. The original quote is from New College MS II fol. 254.
139. Ibid, p. 58.
140. Ibid, p. 59.
141. Newton, *The Chronology of Ancient Kingdoms Amended,* p. 210.
142. Ibid, pp. 186–187.
143. Ibid, p. 187.
144. Ibid, p. 202.
145. Ibid, p. 14.
146. Ibid, p. 16.
147. Ibid, p. 11.
148. Ibid, p. 16.

149. Newton, "A Short Chronicle from the First Memory to the Conquest of Persia by Alexander the Great", p. 16.
150. Newton, *The Chronology of Ancient Kingdoms Amended*, p. 332.
151. Newton "A Short Chronicle from the First Memory to the Conquest of Persia by Alexander the Great", p. 39.
152. This will be examined in Chap. 7.
153. Stukeley, *The Family Memoirs of the Rev William Stukeley MD and the Antiquarians and Other Correspondence of William Stukeley, Roger and Samuel Gales Etc*, p. 262.
154. Such as New College MS 361.1B. Transcripts of these manuscripts are available through the Newton Project at http://www.newtonproject.sussex.ac.uk.
155. See Chap. 7.
156. Stukeley, *Memoirs of Sir Isaac Newton*, p. 18.
157. Ibid, p. 18.
158. See p. 199.
159. See p. 313.
160. Newton, *The Chronology of Ancient Kingdoms Amended*, pp. 212–213.
161. Ibid, p. 221.
162. Ibid, p. 160.
163. Anonymous, "Newton's 'Principia' – Bicentenary of Publication", *The Times* 1887, p. 4.
164. Paola Casini, "Newton: The Classical Scholia", *History of Science* 22 (1984), pp. 1–2.
165. For the text of Gregory's Introduction see pp. 47–57.
166. Niccolò Guicciardini, *Reading the Principia* (Cambridge: Cambridge University Press, 1999) see footnote p. 102; D. T. Whiteside, ed., *The Mathematical Papers of Isaac Newton* Vol. 8 (Cambridge: Cambridge University Press, 1981) p. 459.
167. Ibid, p. 102.
168. Ibid, pp. 102–103; Whiteside, *The Mathematical Papers of Isaac Newton,* p. 453.
169. Table copied from Casini, "Newton: The Classical Scholia", p. 8.
170. My translation from Latin of Newton's draft proposition reprinted in Ibid, pp. 31–32.
171. Frances A. Yates, *Giordano Bruno and the Hermetic Tradition* (London: Routledge, 2002), p. 8.
172. Betty Jo Teeter Dobbs, "Newton's Commentary on the Emerald Tablet of Hermes Trismegistus", in *Its Scientific and Theological Significance, in Hermeticism and Renaissance: Intellectual History and the Occult in Early Modern Europe*, eds. Ingrid Merkel and Allen G Debus (London and Toronto: Associated University Presses, 1988).
173. Isaac Newton, "Appendix A – Of natures obvious laws & processes in vegetation", in *The Janus Faces of Genius,* ed. and trans. B. J. T. Dobbs (Cambridge: Cambridge University Press, 2002), pp. 276–277.
174. Ibid, p. 269.
175. Hermes Trismegist, "Tabula Saragdina", in *The Janus Faces of Genius,* ed. B. J. T. Dobbs and trans. Isaac Newton (Cambridge: Cambridge University Press, 2002), p. 274.
176. Isaac Newton, "Commentarium", in Dobbs, *The Janus Faces of Genius,* pp. 256–270.
177. Isaac Newton, *Notes from Ramon Lull* (unpublished manuscript, Stanford, California: Stanford University Library, undated).
178. Isaac Newton, *Experimenta Raymundi* (unpublished manuscript, Cambridge: King's College, undated).
179. Isaac Newton, *Tabula Smaragdina and Hieroglyphica Planetarum* (unpublished manuscript, Cambridge: King's College, undated).
180. Dobbs, *The Janus Faces of Genius,* p. 180; Westfall, *Never at Rest,* p. 529.
181. Isaac Newton, *Epistola ad verso Hermetis Discipulos continens Claves Sex Principales Philosophiae Secretae, Planetarum* (unpublished manuscript, Cambridge: King's College, early 1690s).
182. Ibid, p. 277.
183. Dobbs, "Newton's Commentary on the Emerald Tablet of Hermes Trismegistus", pp. 184–188.
184. Dobbs, *The Janus Faces of Genius,* pp. 89–212.
185. Paola Casini, "Newton: The Classical Scholia", p. 2.

186. Manuel, *Isaac Newton: Historian*, p. 100.
187. Ibid, p. 17.
188. Isaac Newton, *Fair copies of the "Short Chronicle and Chronology of the Ancient Kingdoms Amended"* (unpublished manuscript, Cambridge: Cambridge University Library, c1725).
189. Dobbs, *The Janus Faces of Genius*, p. 7.

Prophesy and the Temple

190. Philip C. Almond, "Henry More and the Apocalypse", *Journal of the History of Ideas* 54, no. 2 (1993), p. 189; Jeffrey K. Jue, *Heaven Upon Earth: Joseph Mede (1586–1638) and the Legacy of Millenarianism* (Dordrecht: Springer, 2006), pp. 4–6.
191. Jue, *Heaven Upon Earth: Joseph Mede (1586–1638) and the Legacy of Millenarianism,* p. 35.
192. Ibid, pp. 81–82; Sarah Hutton, "More, Newton, and the Language of Biblical Prophecy", in *The Books of Nature and Scripture: Recent Essays on Natural Philosophy, Theology, and Biblical Criticism in the Netherlands of Spinoza's Time and the British Isles of Newton's Time*, eds. James E. Force and Richard H. Popkin (Dordrecht: Kluwer, 1994), p. 42.
193. Hutton, "More, Newton, and the Language of Biblical Prophecy", p. 43.
194. Almond, "Henry More and the Apocalypse, p. 189.
195. Hutton, "More, Newton, and the Language of Biblical Prophecy", p. 44; Sarah Hutton, "Henry More and the Apocalypse", in *Prophecy and Eschatology* ed. Michael Wilks (Oxford: Blackwell, 1994), p. 135.
196. Maurizio Mamiani, "Newton on Prophecy and the Apocalypse", in *The Cambridge Companion to Newton*, eds. I. Bernard Cohen and George E. Smith (Cambridge: Cambridge University Press, 2002), p. 387; Hall, *Isaac Newton: Adventurer in Thought*, pp. 372–373; Michael Murrin, "Newton's Apocalypse", in *Newton and Religion: Context, Nature, and Influence*, eds. James E. Force and Richard H. Popkin (Dordrecht: Kluwer, 1999), see note 32, p. 209.
197. Westfall, *Never at Rest: A Biography of Isaac Newton*, p. 349; Hutton, "More, Newton, and the Language of Biblical Prophecy", p. 41.
198. Florian Cajori, "Sir Isaac Newton's Early Study of the Apocalypse", *Popular Astronomy* XXXIV, no. 2 (1926).
199. Murrin, "Newton's Apocalypse", p. 211.
200. Joseph Mede, *The Key of the Revelation* (London: Printed for Philip Stephen at his Shop in Pauls Church Yard, 1643), Part 2, p. 3.
201. Westfall, *Never at Rest: A Biography of Isaac Newton,* p. 348.
202. Bob Iliffe, " 'Making a Shew:' Apocalyptic Hermeneutics and the Sociology of Christian Idolatry in the Work of Isaac Newton and Henry More", in *The Book of Nature and Scripture: Recent Essays on Natural Philosophy, Theology, and Biblical Criticism in the Netherlands of Spinoza's Time and the British Isles of Newton's Time*, eds. James E. Force and Richard H. Popkin (Dordrecht: Kluwer, 1994), pp. 75–82.
203. Mamiani, "Newton on Prophecy and the Apocalypse", p. 395.
204. Newton, *Yahuda Ms 1.1*, p. 15r.
205. Drawn by the author from Mede, Joseph. *The Key of the Revelation* (London: Printed for 1643) part 2, p. 6.
206. Daniel 12:10.
207. Newton, *Yahuda Ms 1.1,* p. 12r.
208. Newton as quoted by Manuel, *The Religion of Isaac Newton,* p. 120.
209. Goldish, *Judaism in the Theology of Sir Isaac Newton*, pp. 11–12.
210. Hall, *Isaac Newton: Adventurer in Thought*, p. 372.
211. Westfall, *Never at Rest: A Biography of Isaac Newton*, p. 817.

212. James E. Force, *William Whiston* (Cambridge: Cambridge University Press, 1985), p. 99.
213. Ibid, p. 140.
214. Isaac Newton, *Observations on the Prophecies of Daniel and the Apocalypse of St. John* (Lewiston, Queenston & Lampeter: The Edwin Mellen Press, 1999), p. 256.
215. Isaac Newton, *The First Book Concerning the Language of the Prophets (Keynes Ms 5)* (unpublished manuscript, Cambridge: King's College, c mid-1680s), fol. 1.
216. Matania Z Kochavi, "One Prophet Interprets Another: Sir Isaac Newton & Daniel", in *The Books of Nature and Scripture: Recent Essays on Natural Philosophy, Theology, and Biblical Criticism in the Netherlands of Spinoza's Time and the British Isles of Newton's Time*, eds. James E. Force and Richard H. Popkin (Dordrecht: Kluwer, 1994), p. 106; Ralph W. V. Elliott, "Isaac Newton's 'of an Universall Language' ", *The Modern Language Review* 52, no. 1 (1957), p. 3; Ralph W. V. Elliott, "Isaac Newton as Phonetician", *The Modern Language Review* 49, no. 1 (1954), p. 6.
217. Elliott, "Isaac Newton's 'of an Universall Language' ", p. 5.
218. Richard S. Westfall, "Newton's Theological Manuscripts", in *Contemporary Newtonian Research*, ed. Zev Bechler (Dordrecht: D. Reidel Publishing, 1982), p. 135.
219. Newton, *The First Book Concerning the Language of the Prophets (Keynes Ms 5)*, fol. 1.
220. Isaac Newton, *Observations on the Prophecies of Daniel and the Apocalypse of St. John* (Lewiston, Queenston & Lampeter: The Edwin Mellen Press, 1999); Newton, *Yahuda Ms 1.1*; Newton, *The First Book Concerning the Language of the Prophets (Keynes Ms 5)*.
221. Newton, *The First Book Concerning the Language of the Prophets (Keynes Ms 5)*, fol. 1v.
222. Mede, *The Key of the Revelation*, Part 1, pp. 56–57.
223. Newton, *The First Book Concerning the Language of the Prophets (Keynes Ms 5)*, fol. 2. The Chaldaean paraphrase is the paraphrased Aramaic translation of the Hebrew Bible, the *Targum*.
224. Ibid, p. 2.
225. Mamiani, "Newton on Prophecy and the Apocalypse", p. 402.
226. Newton, *Yahuda Ms 1.1*, pp. 20r–23r.
227. Ibid, fol. 15r.
228. Ibid, fol. 12r–20r.
229. Kochavi, "One Prophet Interprets Another: Sir Isaac Newton & Daniel", p. 109.
230. Newton, "Yahuda Ms 1.1", p. 9r.
231. Newton, *Observations on the Prophecies of Daniel and the Apocalypse of St. John*, p. 76.
232. Ibid, p. 84.
233. Ibid, p. 73.
234. Ibid, p. 247.
235. S. J. Barnett, "Introduction", in *Observations on the Prophecies of Daniel and the Apocalypse of St. John*, ed. S. J. Barnett (Lewiston, Queenston & Lampeter: The Edwin Mellen Press, 1999), p. 4.
236. Manuel, *Isaac Newton: Historian*, p. 10.
237. Daniel, 12:4.
238. Newton, *Observations on the Prophecies of Daniel and the Apocalypse of St. John*, p. 255.
239. See p. 151.
240. Newton, *Observations on the Prophecies of Daniel and the Apocalypse of St. John*, p. 259; Isaac Newton, *Irenicum* (unpublished manuscript, Cambridge: King's College, undated), fol. 48.
241. Newton, *The First Book Concerning the Language of the Prophets (Keynes Ms 5)*, fol. v.
242. Ibid, fol. 6v.
243. Isaac Newton, *The Original of Religions (Yahuda Ms 41)* (unpublished manuscript, Jerusalem: Jewish National and University Library, undated), fol. 1r.
244. A Prytanæum is a Greek public building where a sacred fire was kept burning.
245. Newton, *The Original of Religions (Yahuda Ms 41)*, fol. 6r.
246. Ibid, fol. 6v.
247. See p. 153.
248. Mandelbrote, " 'a Duty of the Greatest Moment' Isaac Newton and the Writing of Biblical Criticism", p. 283.

249. Newton, *The Original of Religions (Yahuda Ms 41),* fol. 7r.
250. Newton, *The First Book Concerning the Language of the Prophets (Keynes Ms 5)*, pp. 7–8.
251. Ibid, p. 8.
252. Newton, *Observations on the Prophecies of Daniel and the Apocalypse of St. John*, pp. 261–262.
253. This description comes from Ezekiel 1:10, however in Ezekiel 41:18 the cherubim have two faces and it is the two faced cherubim that Newton used in his reconstruction.
254. Newton, *Observations on the Prophecies of Daniel and the Apocalypse of St. Johns*, p. 262.
255. Genesis 49:17.
256. Newton, *Observations on the Prophecies of Daniel and the Apocalypse of St. John,* p. 262.
257. Ibid, p. 265.
258. Ibid, p. 270.
259. Newton, *The First Book Concerning the Language of the Prophets (Keynes Ms 5),* p. 16.
260. Ibid.
261. Ibid, p. 41.
262. Newton, *Observations on the Prophecies of Daniel and the Apocalypse of St. John*, p. 272.
263. Kubrin, "*Newton and the Cyclical Cosmos: Providence and the Mechanical Philosophy"*, p. 326.
264. Taylor, "*Solar System Evolution"*, p. 8.
265. Kubrin, "*Newton and the Cyclical Cosmos: Providence and the Mechanical Philosophy"*, p. 332.
266. Ibid p. 346; Peter Harrison, "Prophecy, Early Modern Apologetics, and Hume's Argument against Miracles", *Journal of the History of Ideas* 60, no. 2 (1999), p. 253; Ayval Ramati, "The Hidden Truth of Creation: Newton's Method of Fluxions", *The British Journal for the History of Science* 34, no. 4 (2001), pp. 417 and 421.
267. Kubrin, "*Newton and the Cyclical Cosmos: Providence and the Mechanical Philosophy"*, p. 346.
268. Edward Chumney, *The Seven Festivals of the Messiah* (Shippensburg: Destiny Image Publishers, 1994), p. 16.
269. Newton, "*The Chronology of Ancient Kingdoms Amended"*, p. 77.
270. Popkin, "Newton's Biblical Theology and His Theological Physics", p. 91; Ayval Leshem, *Newton on Mathematics and Spiritual Purity* (Dordrecht: Kluwer, 2003), p. 195; Richard H. Popkin, "Newton as a Bible Scholar", in *Essays on the Context, Nature, and Influences of Isaac Newton's Theology*, eds. James E. Force and Richard H. Popkin (Dordrecht: Kluwer, 1990), p. 114.
271. Newton, *Yahuda Ms 1.1,* p. 15r.
272. Ibid, p. 14r, 15r.
273. For sources see pp. 117–120.
274. Westfall, "*Never at Rest: A Biography of Isaac Newton",* p. 346.
275. Leshem, "*Newton on Mathematics and Spiritual Purity"*, Chap. 10.
276. Rene Taylor, "Architecture and Magic: Consideration on the Idea of the Escorial", in *Essay Presented to Rudolf Wittkower*, eds. Douglas Fraser, Howard Hibbard, and Milton J. Lewine (London & New York: Phaidon, 1969).
277. Juan Bautista Villalpando and Hieronymus Prado*, Ezechielem Explanationes Et Apparatus Urbis Hierolymitani Commentariis Et Imaginibus Illustratus* (Romae, 1604), Vol. 2.
278. Tessa Morrison, "Villalpando's Sacred Architecture in the Light of Isaac Newton's Commentary", in *Nexus: Architecture and Mathematics*, ed. Kim Williams (Turin: Kim Williams Books, 2008).
279. William Whiston, "*A Collection of Authentick Records Belonging to the Old and New Testament",* (London: The British Library, 1727), p. 1070.
280. Force, *William Whiston*, p. 21.
281. Ibid, p. 129.
282. Ibid, p. 74.
283. William Whiston, *Six Dissertation* (London: 1734), p. 273.
284. Ibid, p. 331.
285. Ibid, p. 269 (Whiston's emphasis).
286. John Saint Clair, "*Observations on Certain Passages in Daniel and the Apocalypse of John"* (London: 1755), p. 172.

287. Alexander Keith, *Sketch of the Evidence from Prophecy; Containing an Account of Those Prophecies Which Were Distinctly Foretold, and Which Have Been Clearly or Literally Fulfilled. With an Appendix, Extracted from Sir Isaac Newton's Observations on the Prophecies.* (Edinburgh: Waugh & Innes, 1823); Joseph Lomas Towers, *Illustrations of Prophecy: In the Course of Which Are Elucidated Many Predictions Which Occur in Isaiah, Daniel, the Writings of the Evangelists, and of the Book of Revelation* (London: 1796).
288. See p. 175.
289. See p. 165.
290. See p. 193.
291. See p. 193.
292. See pp. 293–297.
293. See pp. 307–309.
294. Isaac Newton, "*Treatise on the Revelation (Yahuda Ms 9.1)*" (unpublished manuscript, Jerusalem: Jewish National and University Library, c mid 1680s).
295. Ibid, fol. 4r.

Prytanæum and the Floor Plan of the Temple of Solomon

296. Isaac Newton, *Rough Draft of and Notes for 'Theologiæ Gentilis Origines Philosophicæ' (Yahuda Ms 16)* (unpublished manuscript, Jerusalem: Jewish National and University Library, c1684).
297. Kenneth J. Knoespel, "Interpretive Strategies in Newton's Theologiæ Gentilis Origines Philosophicæ'", in *Newton and Religion: Context, Nature, and Influence*, eds. James E. Force and Richard H. Popkin (Dordrecht, Boston, and London: Kluwer Academic Publishers, 1999), pp. 180–183.
298. Newton, *The Original of Religions (Yahuda Ms 41).*
299. Newton, *Irenicum.*
300. Newton as quoted by Knoespel, "Interpretive Strategies in Newton's Theologiæ Gentilis Origines Philosophicæ", originally from Yahuda MS 16 p. 1r.
301. Newton as quoted by Ibid, p. 189. Originally from Yahuda MS 16, p. 2r.
302. Newton, "*The Original of Religions (Yahuda MS 41)*", fol. 1r.
303. Ibid, fol. 4v.
304. Ibid, fol. 2r.
305. Ibid, fol. 5r.
306. Ibid, fol. 6r.
307. Ibid, fol. 7r.
308. Richard H. Popkin, "Polytheism, Deism and Newton", in *Essays on the Context, Nature, and Influence of Isaac Newton's Theology*, eds. James E. Force and Richard H. Popkin (Dordrecht: Kluwer Academic Publishers, 1990), p. 28.
309. Harrison, "*The Library of Isaac Newton*", p. 258.
310. C. S. Rademaker, "*Life and Work of Gerardus Joannes Vossius*" (Assen: Van Gorcum, 1981), p. 308.
311. Knoespel, "*Interpretive Strategies in Newton's Theologiæ Gentilis Origines Philosophicæ*", p. 186.
312. Goldish, "*Judaism in the Theology of Sir Isaac Newton*", pp. 50–52.
313. Westfall, "*Isaac Newton's Theologiae Gentilis Origines Philosophicae*", p. 26.
314. Rene Taylor, "Hermetism and Mystical Architecture in the Society of Jesus", in *Baroque Art: The Jesuit Contribution*, eds. Rudolf Wittkower and Irma B Jaffe (New York, NY: Fordham University Press, 1972), p. 74.
315. Morrison, "*Villalpando's Sacred Architecture in the Light of Isaac Newton's Commentary*".

316. Isaac Newton, *Miscellaneous Notes and Extracts on the Temple, the Fathers, Prophecy, Church History, Doctrinal Issues, Ets (Yahuda Ms 14)* (unpublished manuscript, Jerusalem: Jewish National and University Library, undated), p. 32r.
317. Drawn by the author from Villalpando and Prado, *Ezechielem Explanationes Et Apparatus Urbis Hierolymitani Commentariis Et Imaginibus Illustratus*, vol. 2, p. 467.
318. Exodus 28: 17–20.
319. Ptolemy, *The Almagest*, trans. R. Catesby Taliaferro (Chicago, IL: Encyclopaedia Britannica, 1955).
320. Villalpando and Prado, *"Ezechielem Explanationes Et Apparatus Urbis Hierolymitani Commentariis Et Imaginibus Illustratus"*, p. 417.
321. Drawn by the author from Villalpando and Prado, *Ezechielem Explanationes Et Apparatus Urbis Hierolymitani Commentariis Et Imaginibus Illustratus*, vol. 2, p. 470.
322. Drawn by the author from Villalpando and Prado, *Ezechielem Explanationes Et Apparatus Urbis Hierolymitani Commentariis Et Imaginibus Illustratus*, vol. 2, p. 472.
323. Mendell Lewittes, ed., "*The Code of Maimonides (Mishneh Tornh) : Book Eight, the Book of Temple Service"* (New Haven, CT: Yale University Press, 1957).
324. From Zur Shalev, "Sacred Geography, Antiquarianism and Visual Erudition: Benito Arias Montano and the Maps in the Antwerp Polyglot Bible", *Imago Mundi* 55 (2003), p. 64, with kind permission.
325. Helen Rosenau, "*Vision of the Temple: The Image of the Temple of Jerusalem in Judaism and Christianity"* (London: Oresko Books Ltd, 1979), p. 92.
326. Claude Perrault, "Unknown Designs for the 'Temple of Jerusalem' ", in *Essays Presented to Rudolf Wittkower*, eds. Douglas Fraser, Howard Hibbard, and Milton J. Lewine (London: Phaidon, 1969), p. 146.
327. Ibid, p. 152.
328. Villalpando and Prado, *Ezechielem Explanationes Et Apparatus Urbis Hierolymitani Commentariis Et Imaginibus Illustratus*.
329. Drawn by the author from Louis de Compiegne de Veil, R. Mosis Majemonidae de Sacrificiis liber accesserunt Abarbanekis exordium, seu prcemium commentariorum in leviticum: et Majemonidae Tractatus de Consecration Calendarum et de ratione intercalandi, quae ex Heraeo convertit in sermonem Latinum et notis illustravit L. de Comiegne de Veil. (London, 1683), unpaginated.
330. Newton, *Miscellaneous Notes and Extracts on the Temple, the Fathers, Prophecy, Church History, Doctrinal Issues, Ets (Yahuda Ms 14)*, fol. 32r.
331. See p. 189
332. From James Stevens Curl, *The Art and Architecture of Freemasonry* (London: B.T. Batsford Ltd, 1991), p. 89, with kind permission of Professor James Stevens Curl.
333. Drawn by the author from Villalpando and Prado, *Ezechielem Explanationes Et Apparatus Urbis Hierolymitani Commentariis Et Imaginibus Illustratus,* vol. 2, unpaginated.
334. See p. 267.
335. See p. 267.
336. See p. 303.
337. Newton, *Miscellaneous Notes and Extracts on the Temple, the Fathers, Prophecy, Church History, Doctrinal Issues, Ets (Yahuda Ms 14)*, fol. 32r–33v.
338. Newton, "*The Original of Religions (Yahuda Ms 41)"*, fol. 3r.
339. James E. Force, "Newton, the 'Ancients,' and the 'Moderns' ", in *Newton and Religion: Context, Nature, and Influence*, eds. James E. Force and Richard H. Popkin (Dordrecht: Kluwer, 1999), p. 249; Knoespel, "Interpretive Strategies in Newton's Theologiæ Gentilis Origines Philosophicæ", p. 195; Neville Brown, "*Engaging the Cosmos: Astronomy, Philosophy, and Faith"* (Brighton: Sussex Academic Press, 2006), p. 56; Haycock, " 'the Long-Lost Truth': Sir Isaac Newton and the Newtonian Pursuit of Ancient Knowledge", *Studies in History and Philosophy of Science* 35, no. 3 (2004), p. 615.
340. Inigo Jones, "*The Most Notable Antiquity of Great Britain Vulgarly Called Stone-Heng on Salisbury Plain Restored"* (Menston: Scholar Press, 1972).

341. John Webb, "*A Vindication of Stone-Heng Restored*" (London: Printed by R. Davenport for Tho. Bassett, 1665).
342. Hart, "*Art and Magic in the Court of the Stuarts*", p. 53.
343. Ibid, pp. 201–205.
344. Jones, "*The Most Notable Antiquity of Great Britain Vulgarly Called Stone-Heng on Salisbury Plain Restored*", p. 68.
345. Drawn by the author from Inigo Jones, "*The Most Notable Antiquity of Great Britain, Vulgarly Called Stone-Heng on Salisbury Plain*" (London 1655), pp. 58 and 59.
346. Ibid, p. 106.
347. From Vitruvius, *The Ten Books on Architecture*, trans. Morris Hicky Morgan (New York: Dover Publications, 1960), p. 147 with kind permission.
348. Jones, "*The Most Notable Antiquity of Great Britain Vulgarly Called Stone-Heng on Salisbury Plain Restored*", p. 106.
349. Ibid, p. 106.
350. Ibid, p. 105.
351. Ibid, p. 107.
352. William Stukeley, "*Stonehenge, a Temple Restor'd to the British Druids*" (Forgotten Books 2007), Online at ForgottenBooks.org.
353. William Stukeley, "*Stukeley's 'Stonehenge' an Unpublished Manuscript 1721–1724*" (New Haven and London: Yale University Press, 2005).
354. Stukeley, "*Stonehenge, a Temple Restor'd to the British Druids*", pp. 14–15.
355. Isaac Newton, "A Dissertation Upon the Sacred Cubit of the Jews", in "*Miscellaneous Works of John Greaves Professor of Geometry at Oxford*" (Londres: 1737).
356. Juan Bautista Villalpando and Hieronymus Prado, "*Ezechielem ExplanationesEt Apparatus Urbis Hierolymitani Commentariis Et Imaginibus Illustratus*", p. 414.
357. See pp. 189, 225, 241 and 260.
358. See p. 267.
359. See p. 209.
360. Maimonides. *The Code of Maimonides (Mishneh Tornh): Book Eight, the Book of Temple Service*, trans. Mendell Lewittes (New Haven: Yale University Press, 1957), p. 9.
361. Ibid, p. 10.
362. Ibid, p. 11.
363. See p. 191.
364. See p. 193.
365. Maimonides. "*The Code of Maimonides (Mishneh Tornh): Book Eight, the Book of Temple Service*", pp. 12 and 24.
366. Exodus 20:23 as quoted by Ibid, p. 9.
367. Maimonides. "*The Code of Maimonides (Mishneh Tornh): Book Eight, the Book of Temple Service*", p. 24.
368. Dobbs, "*The Janus Faces of Genius*", p. 166.
369. Wisdom 11: 21.

The Temple Measurements and the Sacred Cubit

370. Richard T. Glazebrook, "Standards of Measurement, Their History and Development", *The Proceedings of the Physical Society* 43 (1931), p. 413.
371. Drawn by the author at Piraeus Archaeological Museum in Greece.
372. The metrological relief reference number a000103. Image can be seen at the Ashmolean website http://www.ashmolean.org/ash/faqs/q002/q002008.html.
373. Robert Tavernor, *Smoot's Ear* (New Haven and London: Yale University Press, 2007), p. 22.

374. Reference number BM 10057 and BM 10058. Image can be seen on the British Museum's website. http://www.britishmuseum.org/explore/highlights/highlight_objects/aes/r/rhind_mathematical_papyrus.aspx.
375. R. B. Y. Scott, "The Hebrew Cubit", *Journal of Biblical Literature* 77, no. 3 (1958), p. 207.
376. Ibid, p. 207.
377. Herodotus, "*The Histories*" (Harmondsworth: Penguin Books, 1973).
378. Established in Joshua 3: 4, Numbers 35:5, Ruth 1:16, and Exodus 16:29.
379. Josephus, "Jewish Antiquities", in *The New Complete Works of Josephus*, eds. Paul L. Maier and William Whiston (Grand Rapids: Kregel, 1999), XX, 8,6.
380. William Whiston, "The Jewish Wars", in *The New Complete Works of Josephus*, eds. Paul L. Maier and William Whiston (Grand Rapids: Kregel, 1999), V, 2,3.
381. 1 Kings 7:23.
382. 1 Kings 4:26.
383. 2 Chronicles 4:2.
384. 2 Chronicles 4:5.
385. Zur Shalev, "Measurer of All Things: John Greaves (1602–1652) the Great Pyramid, and Early Modern Metrology" *The Journal of the History of Ideas* 63, no. 4 (2002), p. 556.
386. John Greaves, *Pyramidographia: Or, a Description of the Pyramids in Aegypt.* (London: 1646).
387. Shalev, "Measurer of All Things: John Greaves (1602–1652) the Great Pyramid, and Early Modern Metrology", p. 561.
388. Ibid, p. 567.
389. John Greaves, "*A Discourse of the Romane Foot and Denarius, from Whence the Measures and Weights Used by the Ancients May Be Deduced*" (London: 1647).
390. Ibid, p. 1.
391. Shalev, "Measurer of All Things: John Greaves (1602–1652) the Great Pyramid, and Early Modern Metrology", p. 569.
392. Greaves, "*A Discourse of the Romane Foot and Denarius, from Whence the Measures and Weights Used by the Ancients May Be Deduced*", p. 41.
393. Ezekiel 43:13, and the same ambiguity is in Chapter 40:5.
394. In note (b) of translation.
395. Ibid; this is repeated in Yahuda MS 14, f.1.
396. See note (b) of translation.
397. Newton, "A Dissertation Upon the Sacred Cubit of the Jews".
398. Greaves only published a few papers in his lifetime. His collected work was published over eighty years after his death as Thomas Birch, "*Miscellaneous Works of John Greaves*" (London: Printed by J. Hughs for J. Brindley, 1737).
399. Newton, "*A Dissertation Upon the Sacred Cubit of the Jews*", p. 405.
400. Ibid, p. 408.
401. Ibid, p. 421.
402. Jose Faur, "Newton, Maimonidean", *Review of Rabbinic Judaism* 6, no. 2/3 (2004), p. 218.
403. Ibid, p. 408.
404. Newton, "A Dissertation Upon the Sacred Cubit of the Jews", p. 409.
405. Ibid, p. 413.
406. Ibid, p. 414.
407. Ibid, p. 416.
408. Ibid, p. 417.
409. Ibid, p. 419.
410. In fact Villalpando claimed that a cubit was one and a half Roman feet and he referenced Vitruvius, Villalpando does not distinguish between Roman and Jewish cubits, Newton also claimed the Roman cubit to be one and a half Roman feet however he does distinguish between Roman and Jewish cubits; Villalpando and Prado, *Ezechielem ExplanationesEt Apparatus Urbis Hierolymitani Commentariis Et Imaginibus Illustratus,* p. 430.

411. Ibid, p. 421 Newton's references to the Talmud are incorrect. His reference is Mishnaioth, Tract. De Ghaburim, cap. 4; it should be Talmud, Erubin, 48a.
412. Erubin, 48a.
413. Vitruvius, "*The Ten Books on Architecture*", Book 3, Chapter I.
414. Newton, "*A Dissertation Upon the Sacred Cubit of the Jews*", p. 422.
415. Vitruvius, "*The Ten Books on Architecture*", 3. 1. 3.
416. Josephus, *Jewish Antiquities*, IX vols., vol. VIII (London: William Heinemann Ltd, 1963), .V. 2. 3 and XX.8.6.
417. Newton, "A Dissertation Upon the Sacred Cubit of the Jews", p. 424.
418. Vitruvius, "*The Ten Books on Architecture*", 3.4.4.
419. Drawn by author from Newton's description in Isaac Newton, "A Dissertation Upon the Sacred Cubit of the Jews", in "*Miscellaneous Works of John Greaves Professor of Geometry at Oxford*" (Londres: 1737).
420. Newton, "A Dissertation Upon the Sacred Cubit of the Jews", fol. 27.
421. Ibid, fol. 43.
422. See pp. 221–225.
423. The text "seven palms and three digits will be $2\frac{717}{1000}$" should read "$2\frac{218}{1000}$" as stated on p. 412.
424. Ibid, p. 432.
425. Ibid, p. 405.
426. Ibid, p. 427.
427. Ibid, p. 409.
428. Ibid, p. 418.
429. Ibid, p. 427.
430. Ibid, p. 428.
431. Newton, "A Dissertation Upon the Sacred Cubit of the Jews", pp. 427–429.
432. See p. 219.
433. Ibid, note (b) fol. 43.
434. Ezekiel 43:13.
435. This section is repeated in Isaac Newton, "*Notes on the Temple of Solomon and a Comparison of the Measurement Systems*", (unpublished manuscript, Harry Ransom Humanities Research Centre, University of Texas, Austin, Texas, undated).

An Overview of the Contents and the Source of Babson MS 0434

436. See p. 151.
437. See p. 153.
438. See p. 193.
439. See pp. 155–159.
440. See p. 161.
441. See p. 193.
442. See p. 217.
443. See pp. 227–231.
444. See p. 223.
445. See p. 223.
446. See pp. 189, 225, 241 and 259.
447. See p. 223.
448. See p. 321.
449. See p. 295.
450. See p. 311.
451. Westfall, "*Never at Rest: A Biography of Isaac Newton*", p. 346.
452. Faur, "Newton, Maimonides, and Esoteric Knowledge", p. 528; Jose Faur, "Newton, Maimonidean", *Review of Rabbinic Judaism* 6, no. 2/3 (2004), p. 218.

453. Faur, "Newton, Maimonidean", p. 219.
454. Manuel, "*The Religion of Isaac Newton*", p. 84.
455. Matt Goldish, "*Judaism in the Theology of Sir Isaac Newton*" (London: Kluwer, 1998), p. 18.
456. Harrison, "*The Library of Isaac Newton*", p. 74.
457. Ibid, p. 76, also see catalogue numbers 321 and 322.
458. Newton, "A Dissertation Upon the Sacred Cubit of the Jews", p. 421.
459. Goldish, "*Judaism in the Theology of Sir Isaac Newton*", p. 43.
460. Ibid, p. 20.
461. Harrison, "*The Library of Isaac Newton*", Catalogue numbers 1018 and 1020.
462. The "Tractate Middot" is the part of the Talmud that deals with the architecture of the Temple.
463. See pp. 209 and 239.
464. Newton, "A Dissertation Upon the Sacred Cubit of the Jews".
465. See p. 209.
466. Harrison, "*The Library of Isaac Newton*", catalogue number 216.
467. Wolfgang Herrmann, "Unknown Designs for the Temple of Jerusalem", in *Essays in the History of Architecture*, eds. Douglas Fraser, Howard Hibbard, and Milton J. Lewine (Bath: Pitman Press, 1969), Note 24.
468. Newton, "Introduction to the Lexicon of the Prophets, Part Two: About the Appearance of the Jewish Temple", fol. 29 and 37.

Reconstruction

469. There are similar, but less detailed, diagrams in Yahuda MS 14, fol. 1r–8v; the section is entitled "Ezek 40 &c schematis illustrate"; this appears to predate Babson MS 0434. These diagrams may explain the more detailed floor plan at the beginning of Babson MS 0434.
470. Westfall, "*Never at Rest: A Biography of Isaac Newton*", p. 346; Castillejo, "*The Expanding Force in Newton's Cosmos*", pp. 44–45.
471. See p. 175.
472. Ezekiel 41:11.
473. See p. 281.
474. Ibid.
475. See p. 329.
476. Drawn by the author from Isaac Newton, "*The Chronology of Ancient Kingdoms Amended*" (Dublin, 1782), p. 346.
477. See p. 187.
478. Drawn by the author from Isaac Newton, "*The Chronology of Ancient Kingdoms Amended*" (Dublin, 1782), p. 346.
479. Drawn by the author from Isaac Newton, "*The Chronology of Ancient Kingdoms Amended*" (Dublin, 1782), p. 346.
480. Newton, "*The Chronology of Ancient Kingdoms Amended*", p. 341.
481. See p. 323.
482. Newton, "*The Chronology of Ancient Kingdoms Amended*", p. 342.
483. See p. 311.
484. Newton, "*The Chronology of Ancient Kingdoms Amended*", p. 335.
485. Ibid, p. 333.
486. See Chap. 5.
487. Albert Gallatin Mackey, *The Lexicon of Freemasonry* (Philadelphia: Moss & Brother, 1845), p. 357; George Oliver, *The Antiquities of Freemasonry*, (London: G. and W.B. Whittaker, 1823), p. 366; Robert Freke Gould, "*The History of Freemasonry*" (Edinburgh: Jack, 1885).
488. See p. 171.
489. See p. 277.

490. See pp. 319–321.
491. See p. 175.
492. See p. 287.
493. See p. 289.
494. Newton, "*Observations on the Prophecies of Daniel and the Apocalypse of St. John*", p. 261.
495. See pp. 237–239.
496. See pp. 173–175.
497. See pp. 325–329.
498. Villalpando and Prado, "*Ezechielem ExplanationesEt Apparatus Urbis Hierolymitani Commentariis Et Imaginibus Illustratus*", p. 430.
499. See pp. 315–317.
500. From Isaac Newton, "*Introduction to the Lexicon of the Prophets, Part Two: About the Appearance of the Jewish Temple*", (Babson Ms 0434), (unpublished manuscript, Babson College) fol. 9r, with kind permission of The Huntington Library.
501. Drawn by the author from Isaac Newton's description in Babson Ms 0434.
502. Drawn by the author from Isaac Newton's description in Babson Ms 0434.
503. Drawn by the author from Isaac Newton's description in Babson Ms 0434.
504. Drawn by the author from Isaac Newton's description in Babson Ms 0434.
505. Drawn by the author from Isaac Newton's description in Babson Ms 0434.
506. Drawn by the author from Isaac Newton's description in Babson Ms 0434.
507. Drawn by the author from Isaac Newton's description in Babson Ms 0434.
508. Drawn by the author to describe the path of the coloured images.

Conclusion

509. Keynes, "Newton the Man", p. 368.
510. http://www.newtonproject.sussex.ac.uk.
511. Newton, *De Gravitatione et Aequipondio Fluidorum.*
512. Newton, "Appendix A – Of natures obvious laws & processes in vegetation".
513. Isaac Newton, *The Book of Nicholas Flamel conteining The explication of the Hieroglyphical Figures* w^{ch} *he caused to be put in the Church of the SS. Innocents at Paris*, (unpublished manuscript: Cambridge, Massachusetts: Massachusetts Institute of Technology).
514. Gjertsen, "*The Newton Handbook*", pp. 88–90; Manuel, "*A Portrait of Isaac Newton*", pp. 213–225; White, "*Isaac Newton: The Last Sorcerer*", pp. 222–253.
515. See the Newton Project http://www.newtonproject.sussex.ac.uk.
516. Westfall, "*Never at Rest*", p. 346.

Translation of Babson MS 0434

517. And three cubits in height.
518. It was cedar and Solomon overlaid it with pure gold.
519. Newton uses the word "peribolo" which is Vulgate, Ezekiel 42:7.
520. Cyrus the Great (576 or 590 BC–530 BC), also known as Cyrus II; he was the founder of the Persian Empire under the Achaemenid Dynasty, mentioned in the Bible 2 Chronicles 36: 22–23, Ezra 1:1–2. 1:7–8, 3:7, 4:3 & 5, 5:13.
521. Hecataeus of Miletus (c550–c490 BC). Mentioned in Herodotus 5.36, 125.
522. Simeon the Just, was the High priest during the time of the second temple.
523. No such verse – Ecclesiastes only goes up to Chapter 12.

524. Drawn by the author from Isaac Newton, "*Introduction to the Lexicon of the Prophets, Part Two: About the Appearance of the Jewish Temple*", (Babson Ms 0434), (unpublished manuscript, Babson College) fol. 8r.
525. Drawn by the author from Isaac Newton, "*Introduction to the Lexicon of the Prophets, Part Two: About the Appearance of the Jewish Temple*", (Babson Ms 0434), (unpublished manuscript, Babson College) fol. 9r.
526. Enlargement of Fig. 2a.
527. Ezekiel does not mention the number of steps.
528. Ezekiel claimed that the foundation of the side chambers was six cubits but this makes no sense since the chambers are five cubits wide and the wall another five cubits, but Newton changes this to the length of the chambers; he removes the width of the wall and only counts the three walls of the chambers. This equals one hundred and eighty cubits as the perimeter.
529. In fact **wh** measures19 cubits, 4 (storerooms) + 5 (walkway) +5 (chambers) + 5 (of the other wall); later Newton claimed that the twenty cubits is the fifteen cubits at the side of the temple in the Separate Place plus the five cubits of the thickness of the Temple wall.
530. Newton wrote "front of" but it is clearly marked on the plan at the back of the Separate Place.
531. Newton has written ωhky but this does not equate with the plan described; it should be ωTZπ.
532. These small atria are removed in the later plan.
533. Drawn by the author from Isaac Newton, "*Introduction to the Lexicon of the Prophets, Part Two: About the Appearance of the Jewish Temple*", (Babson Ms 0434), (unpublished manuscript, Babson College) fol. 13r.
534. Hecataeus' description of the second Temple can be seen in Josephus, *Against Apion* 1.198f.
535. Building a facing that is added to a structure such as a wall or building that provides additional support.
536. Tiberius Julius Alexander was probably born in the early reign of Emperor Tiberius (AD-37). His father was Alexander, an Alexandrian Jew who held the office of Alabarch: the meaning of this term is debated but generally considered to denote a senior customs official. The Alexander senior enjoyed Roman citizenship, which was a rare privilege amount the Jews of Alexandria, this privilege was pass on to his sons. Tiberius Julius Alexander's uncle was the prominent philosopher Philo.
537. The soreg was a low wall surrounding the Holy Temple, which served as a boundary. Beyond this point, entry was permitted only to Jews who were not impure through exposure to death. The soreg featured signs, in a number of languages that warned these unauthorized people against entering the area of the Holy Temple. During the period of the Hasmoneans, the Greek invaders made 13 breaches in the soreg in opposition of this prohibition. After the Hasmonean victory, our Sages partially repaired these damaged areas but left small fences as a remembrance to the destruction. They decreed that anyone walking past any of these rebuilt breaches should bow down and thank God for His salvation. Source: (Tractate Midot 2:3).
538. חיל Chajil is the Hebrew word strength. It is used in Joshua 1:14 and 1 Samuel 16:18 for men of strength and valour and in Proverbs 31:10 and Ruth 3:11 for a woman of virtue.
539. Eliezer ben Jose.
540. Synedrium, is a Greek word that means "assembly" and is especially used of judicial or representative assemblies; it is the name by which (or by its Hebrew transcription, *sanhedrin, sanhedrim*) the supreme council and the highest court of justice at Jerusalem are known in New Testament times.
541. In Vitruvius Book III, Chapter III, "The Proportions of Intercolumniations and of Columns" the smallest interval is one and half the width of the base of the column.
542. Vitruvius, Book IV, Chapter III gives the height of a Doric column as six and a half times the diameter of the base without the capital.
543. Vitruvius, Book III, Chapter III, Section 6. Eustyle is a type of temple whose columns had the better proportions according to Vitruvius, from the point of view of the aesthetics and of solidity. The length of its intercolumnios equalled to two diameters and a fourth of the columns, except the head of the subsequent and previous part; that midal is three diameters.

544. Vitruvius Book III, Chapter III, Sections 4–5. This could refer to diastyle – the interval being three times the thickness of the columns or araeostyles – the interval being four times the thickness of the columns. Both are in danger of the architraves breaking if built in marble but areostyles must be built in wood.
545. In the manuscript reads *advero; this* must be a mistake.
546. Drawn by the author from Isaac Newton, *Introduction to the Lexicon of the Prophets, Part Two: About the Appearance of the Jewish Temple,* (Babson Ms 0434), (unpublished manuscript, Babson College) fol. 42r.
547. "Always" added to emphasize the point.
548. Newton wrote "wide" here but must mean length.
549. The number of rooms is not mentioned by Ezekiel.
550. In the manuscript Newton wrote "peccato" (sin, moral offence), clearly an error.
551. The foot note for this verse does not appear in the original text but this could be for verse 19 as suggested by transcription on the Newton Project http://www.newtonproject.sussex.ac.uk.
552. Should be 1 Kings 8:12.
553. To a ratio of 3 to 2.
554. In fact this adds to 98 cubits. Newton has forgotten to write the 2 cubits of the wall between the holy place and the holy of holies but clearly he has added it in.
555. The measurement from the corner to the gate is 87.5 cubits and although Newton specifies 7 columns in this measurement, thus six inter-columns, these inter-columns are more that a "little" greater to the interior inter-columns.
556. In his drawing of the temple (Fig. 7.4) and the brief description that is included in that drawing he stated there are eight rooms along side the temple and a further four on the other side of the walkway. In this second plan there are now a total of fifteen.
557. These room are divided up into smaller chambers. Newton is using the term "room" in the same sense as Josephus.
558. Gemariah the son of Shaphan the scribe, and one of the Levites of the temple in the time of Jehoiakim (Jeremiah 36:10).
559. Hanun was a king of Ammon described in 2 Samuel.
560. Maaseiah, musician under David, captain who aided the restoration of Joash, officer of King Uzziah, son of King Ahaz, man charged by Josiah with repairing the Temple.
561. An instrument string.
562. I Chronicles 26:14. The lot for the East Gate fell to Shelemiah. Then lots were cast for his son Zechariah, a wise counsellor, and the lot for the North Gate fell to him.
563. Parbar – a place apparently connected with the temple, probably a "suburb" as the word is rendered in 2 Kings 23:11; a space between the temple wall and the wall of the court; an open portico into which the chambers of the official persons opened (1 Chronicles 26:18).
564. Jehoiada – The father of Benaiah, who was one of David's chief warriors (2 Samuel. 8:18; 20:23) He was among the foremost of the benefactors of the kingdom, and at his death was buried in the city of David among the kings of Judah (2 Chronicles. 24:15, 16). He is said to have been one hundred and thirty years old.
565. Drawn by the author from Isaac Newton, "*Introduction to the Lexicon of the Prophets, Part Two: About the Appearance of the Jewish Temple",* (Babson Ms 0434), (unpublished manuscript, Babson College) fol. 70r.

Bibliography

Primary Sources

Birch, Thomas. *Miscellaneous Works of John Greaves*, London: Printed by J. Hughs for J. Brindley, 1737.

Compiegne de Veil, Louis. R. Mosis Majemonidae de Sacrificiis liber accesserunt Abarbanekis exordium, seu prcemium commentariorum in leviticum: et Majemonidae Tractatus de Consecration Calendarum et de ratione intercalandi, quae ex Heraeo convertit in sermonem Latinum et notis illustravit L. de Comiegne de Veil, London, 1683.

Compiegne de Veil, Louis. *Pyramidographia: Or, a Description of the Pyramids in Aegypt*, London, 1646.

Greaves, John. *A Discourse of the Romane Foot and Denarius, from Whence the Measures and Weights Used by the Ancients May Be Deduced*, London, 1647.

Hermes Trismegist, "Tabula Saragdina", in *The Janus Faces of Genius*, edited by B. J. T. Dobbs and translated by Isaac Newton, pp. 274, Cambridge: Cambridge University Press, 2002.

Herodotus. *The Histories*, Harmondsworth: Penguin Books, 1973.

Jones, Inigo. *The Most Notable Antiquity of Great Britain Vulgarly Called Stone-Heng on Salisbury Plain Restored*, Menston: Scholar Press, 1972.

Josephus. "Jewish Antiquities", in *The New Complete Works of Josephus*, edited by Paul L. Maier and William Whiston, pp. 49–641, Grand Rapids: Kregel, 1999.

Josephus. "The Jewish Wars", in *The New Complete Works of Josephus*, edited by Paul L. Maier and William Whiston, pp. 670–936, Grand Rapids: Kregel, 1999.

Maimonides. *The Code of Maimonides (Mishneh Tornh): Book Eight, the Book of Temple Service*, translated and edited by Mendell Lewittes, New Haven, CT: Yale University Press, 1957.

Mede, Joseph. *The Key of the Revelation*, London: Printed for Philip Stephen at his Shop in Pauls Church Yard, 1643.

Newton, Isaac. *The Chronology of Ancient Kingdoms Amended*, London: Histories and Mysteries of Man, 1988.

Newton, Isaac. "Commentarium", in *The Janus Faces of Genius*, edited and translated by B. J. T. Dobbs, pp. 276–277, Cambridge: Cambridge University Press, 2002.

Newton, Isaac. *The Book of Nicholas Flamel Conteining the Explication of the Hieroglyphical Figures wch He Caused to be Put in the Church of the SS. Innocents at Paris*, unpublished manuscript, Cambridge, MA: Massachusetts Institute of Technology, undated.

Newton, Isaac. *De Gravitatione et Aequipondio Fluidorum*, unpublished manuscript, Cambridge: Cambridge University Library, undated.

Newton, Isaac. "A Dissertation Upon the Sacred Cubit of the Jews", in *Miscellaneous Works of John Greaves Professor of Geometry at Oxford*, pp. 405–433, Londres, 1737.

Newton, Isaac. *Epistola ad verso Hermetis Discipulos continens Claves Sex Principales Philosophiae Secretae, Planetarum*, unpublished manuscript, Cambridge: King's College, c1680s.

Newton, Isaac. *Experimenta Raymundi*, unpublished manuscript, Cambridge: King's College, c1680s.

Newton, Isaac. *The First Book Concerning the Language of the Prophets*, unpublished manuscript, Cambridge: King's College, c mid-1680s.

Newton, Isaac. *Introduction to the Lexicon of the Prophets, Part Two: About the Appearance of the Jewish Temple*, unpublished manuscript, Wellesley, MA: Babson College, c1680s.

Newton, Isaac. *Irenicum*, unpublished manuscript, Cambridge: King's College, undated.

Newton, Isaac. *Miscellaneous Notes and Extracts on the Temple, the Fathers, Prophecy, Church History, Doctrinal Issues, Ets (Yahuda Ms 14)*, unpublished manuscript, Jerusalem: Jewish National and University Library, undated.

Newton, Isaac. *Notebook Containing Newton's Expense Accounts from 19 March 1659 Through His Cambridge Years*, unpublished manuscript, Cambridge: Trinity College Library, c1659.

Newton, Isaac. *Notes from Ramon Lull*, unpublished manuscript, Stanford, CA: Stanford University Library, undated.

Newton, Isaac. *Observations on the Prophecies of Daniel and the Apocalypse of St. John*, Lewiston, Queenston & Lampeter: The Edwin Mellen Press, 1999.

Newton, Isaac. *Notes on the Temple of Solomon and a Comparison of the Measurement Systems*, unpublished manuscript, Austin, TX: Harry Ransom Humanities Research Centre, University of Texas, undated.

Newton, Isaac. "Of Natures Obvious Laws & Processes in Vegetation"', in *The Janus Faces of Genius*, edited and translated by B. J. T. Dobbs, pp. 276–277, Cambridge: Cambridge University Press, 2002.

Newton, Isaac. "Optics", in *Mathematical Principles of Natural Philosophy*, edited by Florian Cajori, pp. 377–551, Chicago, IL: The University of Chicago, 1952.

Newton, Isaac. *The Original of Religions (Yahuda Ms 41)*, unpublished manuscript, Jerusalem: Jewish National and University Library, undated.

Newton, Isaac. *Rough Draft of and Notes for 'Theologiæ Gentilis Origines Philosophicæ' (Yahuda Ms 16)*, unpublished manuscript, Jerusalem: Jewish National and University Library, c1684.

Newton, Isaac. "A Short Chronicle from the First Memory to the Conquest of Persia by Alexander the Great", in *The Chronology of Ancient Kingdoms Amended*, edited by J. Conduitt, pp. 9–42, London: Histories & Mysteries of Man, 1988.

Newton, Isaac. *Tabula Smaragdina and Hieroglyphica Planetarum*, unpublished manuscript, Cambridge: King's College, undated.

Newton, Isaac. *Treatise on the Revelation (Yahuda Ms 9.1)*, unpublished manuscript, Jerusalem: Jewish National and University Library, c mid 1680s.

Newton, Isaac. *Yahuda Ms 1.1*, unpublished manuscript, Jerusalem: Jewish National and University Library, undated.

Ptolemy. *The Almagest*, translated by. R. Catesby Taliaferro, Chicago, IL: Encyclopedia Britannica, 1955.

Saint Clair, John. *Observations on Certain Passages in Daniel and the Apocalypse of John*, London, 1755.

Stukeley, William. *The Family Memoirs of the Rev William Stukeley MD and the Antiquarians and Other Correspondence of William Stukeley, Roger and Samuel Gales Etc*, Edinburgh: William Blackwood, 1883.

Stukeley, William. *Memoirs of Sir Isaac Newton*, London: Taylor and Francis, 1936.

Stukeley, William. *Stonehenge, a Temple Restor'd to the British Druids*, Forgotten Books, 2007.

Stukeley, William. *Stukeley's 'Stonehenge' an Unpublished Manuscript 1721–1724*, New Haven and London: Yale University Press, 2005.

Towers, Joseph Lomas. *Illustrations of Prophecy: In the Course of Which Are Elucidated Many Predictions Which Occur in Isaiah, Daniel, the Writings of the Evangelists, and of the Book of Revelation*, London, 1796.

Ussher, James. *The Annals of the World*, London: Printed by E. Tyler, for J. Crook and for G. Bedell, 1658.

Villalpando, Juan Bautista, and Hieronymus Prado. *Ezechielem Explanationes Et Apparatus Urbis Hierolymitani Commentariis Et Imaginibus Illustratus*. Romae, 1604.
Vitruvius. *The Ten Books on Architecture*, translated by Morris Hicky Morgan, New York, NY: Dover Publications, 1960.
Webb, John. *A Vindication of Stone-Heng Restored*, London: Printed by R. Davenport for T. Bassett, 1665.
Whiston, William. *A Collection of Authentick Records Belonging to the Old and New Testament*, London: The British Library, 1727.
Whiston, William. *Six Dissertation*, London, 1734.
Whiteside, D. T. ed. *The Mathematical Papers of Isaac Newton*, Vol. 8, Cambridge: Cambridge University Press, 1981.

Secondary Sources

Almond, Philip C. "Henry More and the Apocalypse", *Journal of the History of Ideas* 54(2): 189–200, 1993.
Anonymous. "Art and Literary Gossip", *Manchester Times*, 1876.
Anonymous. "Essays on Chronology; Being a Vindication of the System of Sir Isaac Newton, by a Member of the University, Cambridge", *Gentleman's Magazine* December, 520, 1827.
Anonymous. "General News", *Plymouth and Cornish Advertiser*, 8th June, 2711: 7, 1870.
Anonymous. "Newton's 'Principia' – Bicentenary of Publication", *The Times*, 4, 1887.
Anonymous. "Sir Isaac Newton Tercentenary", *The Times*, col B, 1946.
Anonymous (W. X). "A Critical and Apologetical Dissertation for Sir Isaac Newton's New System of Chronology and Mythology; Design'd Chiefly as an Introduction to a Farther Enquiry and Examination into Its Truth and Certainty", *Present State of the Republick of Letters* 2, September, pp. 210–220, 1728.
Anonymous (W. X). "A Critical and Apologetical Dissertation for Sir Isaac Newton's New System of Chronology and Mythology; Especially as Compared with That of the Late Bishop Cumberland, & C", *Present State of the Republick of Letters* 2, November, pp. 362–372, 1728.
Barnett, S. J. "Introduction", in *Observations on the Prophecies of Daniel and the Apocalypse of St. John*, edited by S. J Barnett, Lewiston, pp. 1–61, Queenston & Lampeter: The Edwin Mellen Press, 1999.
Brewster, David. *Memoirs of the Life, Writing and Discoveries of Isaac Newton*, Edinburgh: Thomas Constable & Co, 1855.
Brown, Neville. *Engaging the Cosmos: Astronomy, Philosophy, and Faith*, Brighton: Sussex Academic Press, 2006.
Cajori, Florian. "Sir Isaac Newton's Early Study of the Apocalypse", *Popular Astronomy* XXXIV (2):75–78, 1926.
Casini, Paola. "Newton: The Classical Scholia", *History of Science* 22: 1–57, 1984.
Castillejo, David. *The Expanding Force in Newton's Cosmos*, Madrid: Ediciones De Arte Y Bibliofilia, 1981.
Ceram, C. W. *Gods, Graves and Scholars: The Story of Archaeology*, London: Sidgwick and Jackson Limited, 1952.
Chittenden, N. W. "The Life of Sir Isaac Newton", in *Newton's Principia: The Mathematical Principles of Natural Philosophy*, edited by Andrew Motte and N. W. Chittenden, pp. 9–61, New York, NY: Daniel Adee, 1848.
Christianson, Gale E. *Isaac Newton*, Oxford: Oxford University Press, 2005.
Chumney, Edward. *The Seven Festivals of the Messiah*, Shippensburg: Destiny Image Publishers, 1994.
Clark, David and Stephen P. H. Clark. *Newton's Tyranny*, New York, NY: W.H. Freeman and Company, 2000.

Conduitt, John. "Memoirs of Sir Isaac Newton", in *Collections for the History of the Town and Soke of Grantham Containing Authentic Memoirs of Sir Isaac Newton*, edited by Edmund Turner, pp. 155–172, London: William Miller, 1806.

Cunningham, Andrew. "How the Principia Got Its Name: Or Taking Natural Philosophy Seriously", *History of Science* 29: 377–392, 1991.

Curl, James Stevens. *The Art and Architecture of Freemasonry*, London: B.T. Batsford Ltd, 1991.

Dobbs, Betty Jo Teeter. *The Janus Faces of Genius*, Cambridge: Cambridge University Press, 2002.

Dobbs, Betty Jo Teeter. "Newton's Commentary on the Emerald Tablet of Hermes Trismegistus", in *Its Scientific and Theological Significance, in Hermeticism and Renaissance: Intellectual History and the Occult in Early Modern Europe*, edited by Ingrid Merkel and Allen G Debus, pp. 182–191, London and Toronto: Associated University Presses, 1988.

Einstein, Albert. "Introduction", in *Opticks*, edited by Albert Einstein, pp. lix–lxi, New York, NY: Dover, 1952.

Elliott, Ralph W. V. "Isaac Newton's 'of an Universall Language'", *The Modern Language Review* 52(1): 1–18, 1957.

Elliott, Ralph W. V. "Isaac Newton as Phonetician", *The Modern Language Review* 49(1): 5–12, 1954.

Fara, Patricia. *Newton: The Making of Genius*, London: Picador, 2002.

Faur, Jose. "Newton, Maimonidean", *Review of Rabbinic Judaism* 6(2/3): 215–249, 2004.

Faur, Jose. "Newton, Maimonides, and Esoteric Knowledge", *Cross Currents* 40(4): 526–540, 1990.

Feingold, Mordechai. "Catalogue of the Libraries of Jerome and John Reynolds", in *Private Libraries in Renaissance England*, edited by R. J. Fehrenbach and E. S. Leedhan-Green, pp. 85–111, Binghamton, NY: Medieval & Renaissance Texts & Studies, 1993.

Force, James E. "Newton, the 'Ancients,' and the 'Moderns'", in *Newton and Religion: Context, Nature, and Influence*, edited by James E. Force and Richard H. Popkin, pp. 237–258, Dordrecht: Kluwer, 1999.

Force, James E. "Newton, the Lord God of Israel and Knowledge of Nature", in *Jewish Christians and Christian Jews*, edited by Richard H. Popkin and G. M. Weiner, pp. 131–158, Dordrect: Kluwer, 1994.

Force, James E. *William Whiston*, Cambridge: Cambridge University Press, 1985.

Gjertsen, Derek. *The Newton Handbook*, London & New York: Routledge & Kegan Paul, 1986.

Glazebrook, Richard T. "Standards of Measurement, Their History and Development", *The Proceedings of the Physical Society* 43: 412–457, 1931.

Goldish, Matt. *Judaism in the Theology of Sir Isaac Newton*, London: Kluwer, 1998.

Goldish, Matt. "Newton on Kabbalah", in *The Books of Nature and Scripture: Recent Essays on Natural Philosophy, Theology, and Biblical Criticism in the Netherlands of Spinoza's Time and the British Isles of Newton's Time*, edited by James E. Force and Richard H. Popkin, pp. 88–103, Dordrecht: Kluwer, 1994.

Gorst, Martin. *Measuring Eternity: The Search for the Beginning of Time*, New York, NY: Broadway Books, 2001.

Gould, Robert Freke. *The History of Freemasonry*, Edinburgh: Jack, 1885.

Guicciardini, Niccolò. *Reading the Principia*, Cambridge: Cambridge University Press, 1999.

Hall, A. Rupert. *Isaac Newton: Adventurer in Thought*, Cambridge: Cambridge University Press, 1992.

Hall, A. Rupert. *Isaac Newton: Eighteenth Century Perspectives*, Oxford: Oxford University Press, 1999.

Hall, A. Rupert and Marie Boas Hall, eds. *Unpublished Papers of Isaac Newton*, Cambridge: Cambridge Press, 1962.

Hall, A. Rupert, and Laura Tilling, eds. *The Correspondence of Isaac Newton, 1718–1727*, London: Cambridge University Press, 1977.

Hamilton, James. *London Lights: The Minds that Moved the City that Shook the World*, London: John Murray, 2008.

Harrison, John R. *The Library of Isaac Newton*, New York, NY: The Cambridge University Press, 1978.

Harrison, Peter. "Prophecy, Early Modern Apologetics, and Hume's Argument against Miracles", *Journal of the History of Ideas* 60(2): 241–256, 1999.

Hart, Vaughan. *Art and Magic in the Court of the Stuarts*, London: Routledge, 1994.

Haycock, David Boyd. "'The Long-Lost Truth': Sir Isaac Newton and the Newtonian Pursuit of Ancient Knowledge", *Studies in History and Philosophy of Science* 35(3): 605–623, 2004.

Hazen, Robert M. *The Diamond Makers*, Cambridge: Cambridge University Press, 1999.

Herrmann, Wolfgang. "Unknown Designs for the Temple of Jerusalem", in *Essays in the History of Architecture*, edited by Douglas Fraser, Howard Hibbard and Milton J. Lewine, pp. 143–158, Bath: Pitman Press, 1969.

Hutton, Sarah. "Henry More and the Apocalypse", in *Prophecy and Eschatology* edited by Michael Wilks, pp. 131–140, Oxford: Blackwell, 1994.

Hutton, Sarah. "More, Newton, and the Language of Biblical Prophecy", in *The Books of Nature and Scripture: Recent Essays on Natural Philosophy, Theology, and Biblical Criticism in the Netherlands of Spinoza's Time and the British Isles of Newton's Time*, edited by James E. Force and Richard H. Popkin, pp. 39–53, Dordrecht: Kluwer, 1994.

Iliffe, Bob. "'Making a Shew:' Apocalyptic Hermeneutics and the Sociology of Christian Idolatry in the Work of Isaac Newton and Henry More", in *The Book of Nature and Scripture: Recent Essays on Natural Philosophy, Theology, and Biblical Criticism in the Netherlands of Spinoza's Time and the British Isles of Newton's Time*, edited by James E. Force and Richard H. Popkin, pp. 55–88, Dordrecht: Kluwer, 1994.

Jue, Jeffrey K. *Heaven Upon Earth: Joseph Mede (1586–1638) and the Legacy of Millenarianism*, Dordrecht: Springer, 2006.

Keith, Alexander. *Sketch of the Evidence from Prophecy; Containing an Account of Those Prophecies Which Were Distinctly Foretold, and Which Have Been Clearly or Literally Fulfilled. With an Appendix, Extracted from Sir Isaac Newton's Observations on the Prophecies*, Edinburgh: Waugh & Innes, 1823.

Keynes, John Maynard. "Newton the Man", in *Essays in Biography*, edited by John Maynard Keynes, pp. 363–74, Cambridge: Cambridge University Press, 1972.

Knoespel, Kenneth J. "Interpretive Strategies in Newton's Theologiæ Gentilis Origines Philosophicæ", in *Newton and Religion: Context, Nature, and Influence*, edited by James E. Force and Richard H. Popkin, pp. 179–202, Dordrecht: Kluwer, 1999.

Knoespel, Kenneth J. "Newtonian the School of Time: The Chronology of the Ancient Kingdoms Amended and the Crisis of the Seventeenth Century Historiography", *The Eighteenth Century* 30(3): 19–41, 1989.

Kochavi, Matania Z. "One Prophet Interprets Another: Sir Isaac Newton & Daniel", in *The Books of Nature and Scripture: Recent Essays on Natural Philosophy, Theology, and Biblical Criticism in the Netherlands of Spinoza's Time and the British Isles of Newton's Time*, edited by James E. Force and Richard H. Popkin, pp. 105–122, Dordrecht: Kluwer, 1994.

Kubrin, David. "Newton and the Cyclical Cosmos: Providence and the Mechanical Philosophy", *Journal of the History of Ideas* 28(3): 325–346, 1967.

Leshem, Ayval. *Newton on Mathematics and Spiritual Purtity*, Dordrecht: Kluwer, 2003.

Lewittes, Mendell, ed. *The Code of Maimonides (Mishneh Tornh): Book Eight, the Book of Temple Service*, New Haven, CT: Yale University Press, 1957.

Mackey, Albert Gallatin. *The Lexicon of Freemasonry*, Philadelphia, PA: Moss & Brother, 1845.

Mamiani, Maurizio. "Newton on Prophecy and the Apocalypse", in *The Cambridge Companion to Newton*, edited by I. Bernard Cohen and George E. Smith, pp. 387–407, Cambridge: Cambridge University Press, 2002.

Mamiani, Maurizio. "The Rhetoric of Certainty: Newton's Method in Science and in the Interpretation of the Apocalypse", in *Persuading Science: The Art of Scientific Rhetoric*, edited by Marcello Pera and William R. Shea, pp. 157–172, USA: Science History Publications, 1991.

Mandelbrote, Scott. "'a Duty of the Greatest Moment' Isaac Newton and the Writing of Biblical Criticism", *The British Journal for the History of Science* 26(3): 281–302, 1993.

Manuel, Frank E. *Isaac Newton: Historian*, Cambridge: Cambridge University Press, 1963.

Manuel, Frank E. *A Portrait of Isaac Newton*, Cambridge, MA: The Belknap Press of Harvard University Press, 1968.

Manuel, Frank E. *The Religion of Isaac Newton*, Oxford: The Clarendon Press, 1974.

Markley, Robert. "Newton, Corruption, and the Tradition of the Universal History", in *Newton and Religion: Context, Nature, and Influence*, edited by James E. Force and Richard H. Popkin, pp. 121–143, Dordrecht: Kluwer, 1999.

Morrison, Tessa. "Villalpando's Sacred Architecture in the Light of Isaac Newton's Commentary", in *Nexus: Architecture and Mathematics*, edited by Kim Williams, pp. 79–91. Turin: Kim Williams Books, 2008.

Munro, J. Arthur R. "Notes on the Text of the Parian Marble II", *The Classical Review* 15(4): 355–361, 1901.

Murrin, Michael. "Newton's Apocalypse", in *Newton and Religion: Context, Nature, and Influence*, edited by James E. Force and Richard H. Popkin, pp. 203–220, Dordrecht: Kluwer, 1999.

Oliver, George. *The Antiquities of Freemasonry*, London: G. and W.B. Whittaker, 1823.

Perrault, Claude. "Unknown Designs for the 'Temple of Jerusalem", in *Essays Presented to Rudolf Wittkower*, edited by Douglas Fraser, Howard Hibbard and Milton J. Lewine, pp. 143–158, London: Phaidon, 1969.

Pfizenmaier, Thomas C. "Was Isaac Newton an Arian", *The Journal of the History of Ideas* 58(1): 57–80, 1997.

Popkin, Richard H. "Newton's Biblical Theology and His Theological Physics", in *Newton's Scientific and Philosophical Legacy*, edited by P. B. Scheurer and G. Debrock, pp. 81–97, Dordrecht: Kluwer, 1988.

Popkin, Richard H. "Newton as a Bible Scholar", in *Essays on the Context, Nature, and Influences of Isaac Newton's Theology*, edited by James E. Force and Richard H. Popkin, pp. 103–118, Dordrecht: Kluwer, 1990a.

Popkin, Richard H. "Polytheism, Deism and Newton", in *Essays on the Context, Nature, and Influence of Isaac Newton's Theology*, edited by James E. Force and Richard H. Popkin, pp. 27–42, Dordrecht: Kluwer, 1990b.

Popkin, Richard H. "Some Further Comments on Newton and Maimonides", in *Essays on the Context, Nature, and Influence of Isaac Newton's Theology*, edited by James E. Force and Richard H. Popkin, pp. 1–6, Dordrecht: Kluwer, 1990c.

Ramati, Ayval. "The Hidden Truth of Creation: Newton's Method of Fluxions", *The British Journal for the History of Science* 34(4): 417–438, 2001.

Rochester, Bishop of. "An Account of Want Related to the Publishing of Sir Isaac Newton's Chronology of Ancient Kingdoms", in *The Chronology of Ancient Kingdom Amended*, pp. 1–14, London, 1770.

Rosenau, Helen. *Vision of the Temple: The Image of the Temple of Jerusalem in Judaism and Christianity*, London: Oresko Books, 1979.

Sarton, George. *Ancient Science through the Golden Age of Greece*, London: Courier Dover Publications, 1993.

Scott, R. B. Y. "The Hebrew Cubit", *Journal of Biblical Literature* 77(3): 205–214, 1958.

Shalev, Zur. "Measurer of All Things: John Greaves (1602–1652) the Great Pyramid, and Early Modern Metrology", *The Journal of the History of Ideas* 63(4): 555–575, 2002.

Shalev, Zur. "Sacred Geography, Antiquarianism and Visual Erudition: Benito Arias Montano and the Maps in the Antwerp Polyglot Bible", *Imago Mundi* 55: 56–80, 2003.

Snobelen, Stephen D. "To Discourse of God: Isaac Newton's Heterodox Theology and His Natural Philosophy", in *Science and Dissent in England, 1688–1945*, edited by Paul Wood, pp. 39–65, London: Ashgate, 2004.

Tavernor, Robert. *Smoot's Ear*, New Haven and London: Yale University Press, 2007.

Taylor, Rene. "Architecture and Magic: Consideration on the Idea of the Escorial", in *Essay Presented to Rudolf Wittkower*, edited by Douglas Fraser, Howard Hibbard and Milton J. Lewine, pp. 81–105, London: Phaidon, 1969.

Taylor, Rene. "Hermetism and Mystical Architecture in the Society of Jesus", in *Baroque Art: The Jesuit Contribution*, edited by Rudolf Wittkower and Irma B. Jaffe, pp. 63–97, New York, NY: Fordham University Press, 1972.

Taylor, Stuart Ross. *Solar System Evolution*, Cambridge: Cambridge Press, 1992.

Tomalin, Claire. *Samuel Pepys: The Unequalled Self*, London: Penguin, 2002.

Turnbull, H. W., ed. *The Correspondence of Isaac Newton, Volume Five, 1694–1709*, Cambridge: Cambridge University Press, 1960.

Turnbull, H. W. *The Correspondence of Isaac Newton, Volume Two, 1676–1687*, Cambridge: Cambridge University Press, 1960.

Urban, Mr. "Remarks on Sir Isaac Newton's Chronology", *Gentleman's Magazine* June: 593–595, 1826.

Westfall, Richard S. "Isaac Newton's Theologiae Gentilis Origines Philosophicae", in *The Secular Mind*, edited by W. Warren Wagar, pp. 15–34, New York, NY: Holmes & Meier, 1982a.

Westfall, Richard S. *The Life of Isaac Newton*, Cambridge: Cambridge University Press, 1993.

Westfall, Richard S. *Never at Rest: A Biography of Isaac Newton*, Cambridge: Cambridge University Press, 1980.

Westfall, Richard S. "Newton's Theological Manuscripts", in *Contemporary Newtonian Research*, edited by Zev Bechler, pp. 129–143, Dordrecht: D. Reidel Publishing, 1982b.

Whiston, William. "An Abstract of the Consultation of Sir Isaac Newton". *Present State of the Republick of Letters* 3, April: 247–285, 1729.

White, Michael. *Isaac Newton: The Last Sorcerer*, London: Fourth Estate, 1998.

Yates, Frances A. *Giordano Bruno and the Hermetic Tradition*, London: Routledge, 2002.

Index

A

Abbé Conti (Abate Conte Antonio Conti), 15–16
Abendana, I., 76
Abraham, 20, 44–45
Agricola, J., 66
Alberti, L. B., 77
Alchemy, 3–5, 13–14, 24–27, 29, 101–102
Alexander the Great, 19, 60, 73, 107, 119, 160, 162
Alexandrian Codex, 75, 152, 154–156, 173
Alhokm, I. A., 66
Almagest, 48
Anaxagoras, 23
Apocalypse, 7, 17, 29, 31, 34–40, 62, 73, 86, 103, 105
Apollodor, 15
Aristotle, 24
Arius, 6
Ark of the Covenant, 38, 97, 105
Arundel, Earl of, 15, 19
Ashmolean Museum, 63
Athanasius, 6
Atrium of Israel, 107, 125–126, 130, 132, 150
Atrium of the Priests, 105–108, 116, 125, 145–146, 148–149, 151, 154
Atrium of the Women, 74, 107, 123, 125, 129–132, 136, 150

B

Barbaro, D., 77
Biblia Sacra Polyglotta, 7, 76
Bochart, S., 45
Brewster, D., 1
Bruno, G., 6
Buxtorf, J., 75, 128

C

Cabbal, 6
Cìlum, 58
Calculus, 4, 13, 16
Cappel, L., 51, 59, 75–77, 124, 130, 156
Castillejo, D., 18
Cavelier, G., 16
Champollion, J.-F., 15
Cherabim, 58
Clockwork universe, 5, 38
The Code of Maimonides, 51, 53, 60
Columella, L. I. M., 66
Compiègne de Veil, L., 51, 53, 167
Composite, Columns, 21
Conduitt, C., 3
Conduitt, J., 12, 17–18, 21, 27, 36, 41, 83, 161
Contra Apionem (Against Apion), 75, 124, 126, 131
Corinthian, Columns, 21–22, 121, 123, 135
Cosmological man, 50, 54
Council of Nicaea, 6
Cumberland, Bishop, 16, 83
Cunningham, A., 9
Cyrus, the Great, 73, 107, 136

D

Daniel, Book of, 13, 17, 27, 29, 31–34, 38, 40–41, 77, 103
De architectura, 57, 77
De Gravitatione et Aequipondio Fluidorum, 9, 102
Democritus, 23–24
De theologia gentili, 45
Didier, L. de, 26
Diodorus, 15, 45, 65
A discourse of the Romane foot and denarius, 65
Dissertation, 11, 59, 66, 70–71, 76–77

Dobbs, B. J. T., 25–26
Domitian, Emperor of Rome, 34
Doric, Columns, 21–22, 133–135
Drusius, 76, 94

E
Einstein, A., 1
Emerald Tablet, 24–26
Erubin, Book of, 68
Eusebius, 24, 45
Eustyli, 134
Exodus, 20, 60, 73, 106
Ezechielem Explanationes, 39, 46, 51, 76–77, 101

F
Faur, J., 3, 66, 76
Ficino, 6
Flamsteed, 16
Fréret, N., 15–16

G
Gales, R., 17
Gate of the Ardor, 124
Gate of the Fire, 124
Gate of Jechoniae, 124
Gate of the Offering, 124–125, 132–133
Gate of Parbar, 81
Gate of the Projection, 124, 129–130, 133
Gates of Assupim, 81
Gate of Shallecheth, 81, 150–151
General Scholium, 5–6, 8–9, 26
Goldish, M., 76
Gravitation, 2, 5, 13, 23–24
Gravity, 1, 2, 23–24
Great Atrium, 83, 93, 106–107, 122, 124–125, 131–132, 136, 145–148, 150–152
Great Synedrium, 125
Greaves, J., 65

H
Harmonic proportions, 46
Harmonic ratios, 23, 50, 59
Harrison, J., 13
Haydon, B. R., 2
Hecataeus, 73–74, 119, 126, 131
Herculaneum, 15
Hermes, T., 24–26
Hermetic philosophy, 26, 65
Herodotus, 15, 20, 45, 64–66
Hesychius of Alexandria, 66
Holy of Holies, 83, 96–97, 123, 140
Hooke, R., 13
House of the Ardor, 125

I
Interior Atrium, 84, 106–108, 110–111, 113, 115, 117, 121, 124–125, 129, 132, 134–135, 138–139, 142–143, 145–146, 148, 150, 152, 155–156
Inverse-square law, 1, 23–24
Ionic, Columns, 21
Irenaeus, 33
Isaiah, Book of, 32, 108, 153, 166

J
Jehoash, 151
Jehoiada, 151
Jewish Antiquities, 64, 112, 116, 120–121, 136, 169–170
Jewish Wars, 64, 75, 120–121
John the Divine, 27, 33, 62, 86, 103
Jones, I., 11, 56–59
Jose, E. b, 157–158, 169–170
Josephus, 6, 15, 35, 45–46, 64, 66, 68–71, 74–76, 112, 116, 120, 133–136, 146, 154, 169–170, 173–174

K
Keill, J., 15
Keynes, J. M., 3–4, 6–7, 24, 101
The Key of the Revelation, 29
Knoespel, K. J., 45
Kochavi, M. Z., 33

L
Lamy, B., 51
Leibniz, G. W., 16
L'Empereur, Constantinus, 51, 54, 75–76, 124, 133
Lightfoot, J., 51
Logicae artis compendium, 7–8
Lull, R., 26

M
Macrobius, 24
Maimonides, M., 6–7, 11, 45, 51, 53, 59–61, 74–76, 101, 125–126, 133, 150, 158, 167–168
Mamiani, M., 7
Manuel, F. E., 76
Marsham, J., 45
Master of the Mint, 13, 18
Mechanical universe, 8, 38
Mede, J., 10, 19, 21, 27, 29–32, 40, 56
Metropolitan Museum, 63
Microcosm, 11, 36, 46–48, 50–51, 53–56, 59, 61, 101
Middoth, 51, 75–76, 124, 133
Mirandola, G. P. D., 6
Misheh Torah, 7
Montano, B. A., 51–52, 59, 75, 124, 128
More, J., 10, 29–30

N
Newenham, F., 2
Newtonian Man, 69
Noah, 10–11, 20, 23, 26, 35, 38, 44–45, 61, 76

O
Observations upon the Prophecies of Daniel and the Apocalypse, 17, 31, 40, 164–165, 172
Opticks, 2–3, 5, 8–9, 13, 38
Original of Religions, 35, 43

P
Paetus, P. C. T., 66
Pantheon, 65
Parian Marble, 15, 19
Pellett, T., 3, 17
Pertius, 65
Philandrier, G., 58, 65–66
Philip II, King of Spain, 46
Philo, 6, 74–76, 120–121
Piraeus Archaeological Museum, 63–64
Pitcaire, A., 14
Pliny, 24
Plutarch, 15, 24
Polybius, 65
Portsmouth Collection, 3, 21
Prado, J., 46
Principia, 2–3, 5, 7–8, 13, 18, 22–23, 25–26, 40, 61–62, 102
Prisca Sapientia, 10, 13–14, 16, 18, 20, 22–27
Prolegomena of the Biblia, 76
Prytanìa, 10, 35–38, 43–45
Prytanìum, 10–11, 36, 43–61, 102
Ptolemy, 46, 48
Purchas, S., 66–67
Pyramidographia, 65
Pythagoras, 23–24, 26

R
Raleigh, W., 14
Rhind Mathematical Papyrus, 63
Rochester, Bishop, 18–19
Room of the Bathe, 132
Room of End, 125
Room of Hipparvae, 132
Room of the Utensils of the Cooks, 125
Royal Society, 3, 4, 16, 18
Rudbeck, O., 14

S
Sanderson, R., 7–8
Seraphim, 37
Serlio, S., 77
Short Chronicle, 14–17, 19–21
Simon, 107–108, 122
Sleidan, J., 13
Smith, B., 17, 31, 41
Snellius, W., 66
Souciet, E., 15, 17
St Clair, J., 40
St Jerome, 48, 66, 152–154, 156, 159
Stonehenge, 11, 56–59
Strabo, 15, 45, 66
Stukeley, W., 1–2, 13, 17, 21–22, 59
Suetonius, 65

T
Tabernacle, 11, 30, 35–36, 44–45, 47–49, 55–56, 73, 80, 94, 105–106, 126, 130, 144–145
Temple of Herod, 34, 76, 145–147
Temple Mount, 61, 95
Theodotus, 15

Theologiǐ Gentilis Origines Philosophicǐ, 43
Treatise on Revelation, 41
Tuscan, Columns, 55–56

U
Of an Universall Language, 32
Ussher, J., 14, 17

V
Vestal Temples, 56
Villalpando, J. B., 11, 39, 46–56, 59, 65–66, 68, 75–77
Vitruvian man, 68, 77
Vitruvius, 49, 57–59, 63, 65–66, 68–69, 76–77, 134
Vossius, 6–7, 45

W
Warden of the Mint, 18
Webb, J., 56
Westfall, R., 31, 39, 76
Whiston, W., 16, 29, 31, 39–40
White, M., 1

Z
Zerubbabel, 37–38, 73–74, 107, 136, 145

MIX
Papier aus verantwortungsvollen Quellen
Paper from responsible sources
FSC® C105338

If you have any concerns about our products,
you can contact us on
ProductSafety@springernature.com

In case Publisher is established outside the EU,
the EU authorized representative is:
Springer Nature Customer Service Center GmbH
Europaplatz 3, 69115 Heidelberg, Germany

Printed by Libri Plureos GmbH
in Hamburg, Germany